Robotics Through Scie

Robotics Through Science Fiction

Artificial Intelligence Explained Through Six Classic Robot Short Stories

edited by Robin R. Murphy

The MIT Press
Cambridge, Massachusetts
London, England

This book was set in ITC Stone Sans Std and ITC Stone Serif Std by Toppan Best-set Premedia Limited. Printed and bound in the United States of America.

Library of Congress Cataloging-in-Publication Data

Names: Murphy, Robin, 1957- editor.
Title: Robotics through science fiction : artificial intelligence explained through six classic robot short stories / edited by Robin R. Murphy.
Description: Cambridge, MA : The MIT Press, 2018. | Includes bibliographical references and index.
Identifiers: LCCN 2018011905 | ISBN 9780262536264 (pbk. : alk. paper)
Subjects: LCSH: Science fiction, American. | Robots--Fiction. | Robotics--Fiction. | Artificial intelligence. | Robots in literature.
Classification: LCC PS648.S3 R628 2018 | DDC 813/.0876208--dc23 LC record available at https://lccn.loc.gov/2018011905

10 9 8 7 6 5 4 3 2 1

To my dad, Fred Roberson, for introducing me to science fiction, and to Chris Trowell for being a creative teacher

Contents

Introduction

This book presents six classic science fiction stories with commentary as a means of answering two modern-day questions: *How are intelligent robots programmed?* and *What are the limits of autonomous robots?* Even though the stories were written before 1973, each one illustrates one or more key principles or algorithms from artificial intelligence (AI) that are actually used in robotics today. This book thus serves as a standalone, nontechnical introduction to robotics for anyone who is interested in AI. The stories can also be treated as a set of admittedly fanciful case studies, which serve as a companion to my textbook *Introduction to AI Robotics*. That textbook is written for advanced undergraduate and graduate students in computer science and engineering who do not necessarily have a background in artificial intelligence.

The six short stories are organized in a sequence based on the organization of intelligence in robot systems, with the material in each chapter building on the previous chapters. The book concludes with a summary chapter fusing the lessons learned in each story into a holistic perspective on programming an intelligent robot, including a discussion on the lack of applicability of Asimov's *Three Laws of Robotics*. The stories and their commentary can be read in any order, but some of the terminology used in the discussions may be confusing if the chapters are read out of order.

Chapter 1 explores *telepresence* through Isaac Asimov's story "Stranger in Paradise." This is a good starting point because telepresence is often derided as a degenerative case of programming artificial intelligence for robotics, as the human is working through the robot. Telepresence is actually the appropriate choice for certain tasks and can require remarkably sophisticated algorithms to enable the human and robot to work together effectively. Chapter 2 addresses *behavior-based robotics* using another Asimov story, "Runaround," and illustrates the common programming approach of *potential fields*. Unlike telepresence, behavior-based robots can operate autonomously but are generally only as intelligent as an insect or animal. *Deliberation*, or what most people would consider cognitive intelligence, is introduced in chapter 3 through Vernor Vinge's story "Long Shot." One reason that AI robots are still rare in practice is because they are difficult to test. *Testing* is discussed in chapter 4 with Asimov's "Catch That Rabbit." The discussion also presents the *levels of initiative* framework for thinking about autonomy. Within the past decade, artificial intelligence has expanded beyond making sure a robot works at all to making sure the robot works with and around people, which is called *human-robot interaction*. This interaction is the subject of chapter 5 and Brian Aldiss's story "Supertoys Last All Summer Long," which illustrates the challenges of *natural language understanding*, given how hard it is for even humans to communicate with each other, much less comprehend the *beliefs*, *desires*, and *intentions* of another agent. It also serves as a discussion of the *uncanny valley*. Chapter 6 uses Philip K. Dick's grim classic story "Second Variety" to consider whether autonomy and *machine learning* will allow robots to mutate and begin to violate *ethics*, threatening all of mankind, or if *bounded rationality* will serve as a limit. Chapter 7 provides a summary and review arranged around three questions: How are intelligent robots programmed? What are the limits of autonomous robots? and What topics are missing in this book?

The short stories are also arranged in a human-centric sandwich, starting and ending with the human with technical details in the middle. The first story, "Stranger in Paradise," offers a glimpse of how robots can make our lives better, while the last story, "Second Variety," imagines how robots can make our lives much worse. The fifth story in the collection, "Supertoys Last All Summer Long," gives a glimpse of human-centric design from a robot's viewpoint. These three stories provide an emotional exploration of robotics and serve as the easy-to-hold bread on how robots interact with people. The second and fourth stories are amusing entries featuring Asimov's pair of smart-aleck troubleshooters, while the third story is a more science-oriented piece. These three middle stories are the "meat" about robots sent off to perform dirty, dangerous, or dull tasks. There is little interaction with humans in these stories. This 1:1 ratio of stories about interaction to stories about tasks mirrors research trends. Robotics research efforts have historically concentrated on expanding what tasks robots can do, but recently there has been increasing emphasis on, and apprehension of, how robots work with and around people.

Each chapter is divided into four sections with references. The first is the introduction, with "As You Read the Story" pointers. This section provides an overall summary of the key AI robotics principle that is captured by the story, priming the reader to be on the alert for how the story illustrates the theory. The second section is the story itself, followed by "After You Have Read the Story." This section discusses the story, with spoilers, and whether the implementation described in the story was consistent with theory. It also brings in discussions of secondary principles or topics in AI robotics that were touched on in the story. This third section includes a list of references to the source material from my textbook and a small set of other papers or books for further reading. Each chapter ends with a subjective overall "Reality Score" rating how realistic the story is. The sections surrounding each story are intended to be sufficient for a layman to understand AI robotics on a high level without oversimplifying the material and will hopefully encourage further in-depth study.

The previous paragraphs explain the purpose and the organization of the book, but they don't really explain why there is a book on artificial intelligence for robotics that uses science fiction stories as examples. That "why" takes a bit longer to explain.

Artificial intelligence for robotics can be difficult to teach, especially if there is not a hands-on component to give some tangible example of how systems are actually built. Computer science curriculums are often so constrained that they cannot support a class with an additional credit hour that would permit a lab component. Even if a class has robotics lab time, there are still problems. One is that robots break. Another is that the "fun" artificial intelligence algorithms can be challenging to program and overwhelm some students. A third problem is that hands-on robotics lab time does not scale to teaching large classes or massive online open courses (MOOCs).

My solution for a standard lecture-only class has been to try to work in a few small, hands-on robotics projects on a specific algorithm, and then supplement the projects with case studies of real systems working on a larger scale. The idea is that even if the robots break or there is no time for a lab, the students can have a tangible system that illustrates how intelligent robots actually apply AI theory in practice in order to function. Reading about a system isn't as effective as programming a system in learning how to translate theory into practice, but it is literally about 150 hours less work for a student. I have always liked case studies, having coedited *Artificial Intelligence and Mobile Robots: Case Studies of Successful Robot Systems* in 1998.[1] And that book has a connection to science fiction as well, as it was cited by Michael Crichton in his book *Prey*.[2]

In 2014, I received a small grant from Texas A&M as a Faculty Fellow for Innovation in High-Impact Learning Experiences to formally prepare a set of case studies taken from the domain of disaster robotics, which is my area of research. My choice of case studies was motivated, of course, by my familiarity with this domain but also by the fact that disaster robotics would represent a category of applications that have a positive societal impact. I thought this societal benefit might prevent side issues, such as the debate over the weaponization of drones, from detracting from how the robot programmers applied theory to practice.

As I taught the class, I proudly noticed that the case studies I had selected were well received and the discussions productive and engaging. Then during the last week of class, I was stumped because I couldn't find any case studies in disasters for centralized control architectures. Out of desperation, I brought up the droids in *Star Wars: The Phantom Menace* as an example of centralized versus distributed control and why you shouldn't have all of your robots enslaved by a single controller. The discussions were really productive, wildly engaging, and *fun*. The students worked much harder to enthusiastically express all the ways centralized control was wrong for the application while using the appropriate AI and robotics terminology and references. Everyone, even the guy who slept in the back of the room for the entire semester, had something to contribute or a question to ask. If nothing else, my class walked away from the course with the vocabulary for critiquing popular science fiction—which might actually serve them well in critiquing the futuristic designs, products, and programs they are likely to create and manage in their careers.

Afterward, I remembered how when I first started teaching robotics, each week for extra credit I would pass out a science fiction robot story that was vaguely related to the topic we had covered. The stories were a reward to the students, a chance to reinforce the notion of lifelong reading for their general edification (always emphasized in education), and just kind of fun. Most of the stories were by Isaac Asimov, which was easy to justify because, let's face it, it's hard to be a roboticist without having some exposure to his "I, Robot" stories and Three Laws of Robotics.

I had gotten the idea for giving science fiction robot stories as extra credit from a class I had taken in junior college. Mr. Trowell's anthropology class required three textbooks: a regular textbook that I don't recall, the classic spoof of anthropology "Body Ritual among the Nacirema" (Nacirema is "American" spelled backward), and *Anthropology Through Science Fiction*.[3] Although my parents were probably unhappy about having to pay for three textbooks, it is a tribute to Mr. Trowell's judgment that after thirty years, I still remember the name of the two "quirky" texts and not the Important, Serious Textbook. I had been into science fiction since I was a little girl, reading "Dick and Jane" in school by day and Robert Heinlein's *The Green Hills of Earth* anthology at night,[4] so *Anthropology Through Science Fiction* was a delight. Each science fiction story it contained was already a classic story in its own right, many of which I had already read, but it also clearly and succinctly illustrated some major concept from anthropology. As a new professor in computer science, I adapted Mr. Trowell's strategy. The students in my robotics classes have generally liked the stories and it has provided something a bit different to talk about. I pitched the idea of *Robotics Through Science Fiction* to my editor at the MIT Press in 2004 but they weren't interested at that time and I shifted my attention to creating laboratory exercises.

As a result of the *Star Wars* discussion, I decided to return to my trove of science fiction robot stories that were enjoyable in their own right and also illustrated a key principle in artificial intelligence for robotics. I could have focused on robot movies, which I am doing as a parallel project, but I wanted to preserve the pedagogical principle of encouraging reading. I went back through my original list, racked my memory for what I had been reading in the last two decades, and found and read numerous more recent stories. I settled on the six in this anthology.

The best stories seemed to be the earlier ones, written before the 1980s when robots became more common in the workplace and in popular science. Of course, writers hadn't stopped incorporating robots into plots after Karel Čapek created the concept in his 1921 play *R.U.R.*,[5] and their creativity didn't end in the 1980s, but it seems the earlier stories focused on robot technology and what it means to be a robot, and when robots actually became commonplace in factories, the stories became explorations of what it means to human.

To borrow a term from cinema, the "MacGuffin" in early science fiction stories is some quirk about how robots might work. A MacGuffin is something that sets the plot in motion but isn't necessarily the point of the story. For example, the statue in *The Maltese Falcon* is a MacGuffin; the characters are all pursuing the statue, but the movie is really about the alliances and betrayals between them and we learn very little about the statue. In science fiction, a MacGuffin is often a technological innovation that initiates the plot of the story, be it a quest, overcoming some challenge, or even comedy, but its technological plausibility is more essential to enjoying the story than in regular fiction. In *The Maltese Falcon*, a diamond bracelet or sacred text could have been substituted for the statue without changing the plot, dialog, or impact of the movie. In science fiction stories, the technological MacGuffin cannot be so easily separated from the story. The science is as important as the fiction. Each of the six stories in this book features a MacGuffin that is not only plausible but an essential feature of a truly intelligent robot.

I hope you will enjoy the stories but, more importantly, that you will reach a better understanding of what artificial intelligence is for robotics and what it can really do—and not do. Science fiction gets much of artificial intelligence and robotics wrong, but it is a starting point to discuss what real robots are capable of now and are likely to be able to do in the future.

Let me conclude with a sincere thank-you to the Center for Teaching Excellence, the Office of the Dean of Faculties and Associate Provost, the Office of the Associate Provost for Undergraduate Studies, the Office of the Associate Provost for Graduate and Professional Studies, and the Office of Vice Provost for Academic Affairs at Texas A&M for awarding me an Innovation in High-Impact Learning Experiences fellowship and thus making this book possible. I also deeply appreciate the suggestions from Jim Hendler, Dylan Shell, Bruce Gooch, Craig Marianno, and the anonymous reviewers who devoted their time to critiquing the choice of stories as well as the organization and accuracy of the commentary, the work of Hina Sohrab and Christine Savage in tracking down permissions, Ginny Crossman for her detailed copyediting, and my editor Marie Lufkin Lee.

Robin R. Murphy
2018

1 Telesystems: "Stranger in Paradise"

"Stranger in Paradise" explores *teleoperation* or *telesystems*, a common mechanism by which a user ("operator") can control a robot that is some distance away ("tele"). Along the way the story offers some insights into the motivation for behavior-based robotics and cognitive science approaches to artificial intelligence. The story was written by Isaac Asimov and was published in 1973.[1] It is not in the Three Laws of Robotics canon, so many Asimov aficionados may not be familiar with it.

In the story, one of two biological brothers, Anthony, is working on the Mercury Project. The project goal is to explore that small planet with a robot. The technological MacGuffin for this story is that the robot cannot be successfully teleoperated from Earth as the human operators cannot cognitively adjust to the communication delays. The project managers investigate what would now be called cognitive science or cognitive neuroscience for insights in how to program an autonomous robot that could duplicate human intelligence. The cognitive scientist who joins the project is Anthony's brother William, who has specialized in autism. William's addition to the team leads to rather contrived dramatic tension between the two brothers. Annoying literary devices aside, the story is an excellent primer on telesystems and one of only a handful of stories that can bring me to tears every time I read it.

As You Read the Story ...

The story explicitly discusses teleoperation but is really about telepresence, though the distinction is subtle. The entire field studying how users control robots that are acting out of sight is now generally referred to either as *telesystems*, emphasizing the "tele" or "at a distance" aspect, or as *human supervisory control*, emphasizing that there is a person in the loop. These terms were defined by the expert in the field, Thomas Sheridan, to describe a situation in which one or more human operators is intermittently *giving directives* and *continually receiving information from a computer*, creating an autonomous control loop through artificial effectors and sensors to the controlled process or task environment.[2] If the human operator has sufficient information from the computer such that they feel like they are the robot—essentially virtual reality—then that is generally called telepresence.

Telesystems got their start during the Manhattan Project because scientists could not work directly with radioactive materials, but major advances were made by the space program. It wasn't feasible or cost-effective to have astronauts do everything in space, but people on the ground were stymied by communication delays and the inability to completely see what their robots were doing. During the 1980s telesystems matured as a scientific discipline with formal methods, taxonomies, and subdisciplines, moving beyond the try-it-and-see-if-it-works early days.

Telesystems have been used to take over control of Mars rovers, and the Robonaut astronaut assistant on the International Space Station can be teleoperated. On Earth, the da Vinci medical robot is an example of teleoperation allowing a doctor to work on the "innerspace" of a person. Every ground, aerial, and marine robot used for disasters from 2001 to 2013 was teleoperated even if it had an autonomous option.[3] This included robots that searched deep into the rubble at the 9/11 World Trade Center, flew over areas damaged by Hurricane Katrina, and dove underwater to inspect ports in the wake of the March 2011 Japanese tsunami. The robots used during the Fukushima nuclear accident were teleoperated, and the new generation of robots being used for its decommissioning are also teleoperated due to the complexity of the tasks and environments involved.

As you read the following story, try to identify the project managers' motivation for using a telesystem, as well as the components of the telesystem. The concluding section will give a short history of telesystems and present the formal guidelines that engineers use in deciding when to design a telesystem robot versus a robot that acts independently of an operator.

"Stranger in Paradise" by Isaac Asimov, 1973

1.

They were brothers. Not in the sense that they were both human beings, or that they were fellow children of a creche. Not at all! They were brothers in the actual biological sense of the word. They were kin, to use a term that had grown faintly archaic even centuries before, prior to the Catastrophe, when that tribal phenomenon, the family, still had some validity.

How embarrassing it was! Over the years since childhood, Anthony had almost forgotten. There were times when he hadn't given it even the slightest thought for months at a time. But now, ever since he had been inextricably thrown together with William, he had found himself living through an agonizing time.

It might not have been so bad if circumstances had made it obvious all along; if, as in the pre-Catastrophe days – Anthony had at one time been a great reader of history – they had shared the second name and in that way alone flaunted the relationship.

Nowadays, of course, one adopted one's second name to suit oneself and changed it as often as desired. After all, the symbol chain was what really counted, and that was encoded and made yours from birth.

William called himself Anti-Aut. That was what he insisted on with a kind of sober professionalism. His own business, surely, but what an advertisement of personal poor taste. Anthony had decided on Smith when he had turned thirteen and had never had the impulse to change it. It was simple, easily spelled, and quite distinctive, since he had never met anyone else who had chosen that name. It was once very common – among the pre-Cats – which explained its rareness now perhaps.

But the difference in names meant nothing when the two were together. They looked alike.

If they had been twins – but then one of a pair of twin-fertilized ova was never allowed to come to term. It was just that physical similarity occasionally happened in the non-twin situation, especially when the relationship was on both sides. Anthony Smith was five years younger, but both had the beaky nose, the heavy eyelids, the just noticeable cleft in the chin – that damned luck of the genetic draw. It was just asking for it when, out of some passion for monotony, parents repeated.

At first, now that they were together, they drew that startled glance followed by an elaborate silence. Anthony tried to ignore the matter, but out of sheer perversity – or perversion – William was as likely as not to say. "We're brothers ..."

"Oh?" the other would say, hanging in there for just a moment as though he wanted to ask if they were full blood brothers. And then good manners would win the day and he would turn away as though it were a matter of no interest. That happened only rarely, of course. Most of the people in the Project knew – how could it be prevented? – and avoided the situation.

Not that William was a bad fellow. Not at all. If he hadn't been Anthony's brother; or if they had been, but looked sufficiently different to be able to mask the fact, they would have gotten along famously.

As it was – It didn't make it easier that they had played together as youngsters, and had shared the earlier stages of education in the same creche through some successful maneuvering on the part of Mother. Having borne two sons by the same father and having, in this fashion, reached her limit (for she had not fulfilled the stringent requirements for a third), she conceived the notion of being able to visit both at a single trip. She was a strange woman.

William had left the creche first, naturally, since he was the elder. He had gone into science – genetic engineering. Anthony had heard that, while he was still in the creche, through a letter from his mother. He was old enough by then to speak firmly to the matron, and those letters stopped. But he always remembered the last one for the agony of shame it had brought him.

Anthony had eventually entered science, too. He had shown talent in that direction and had been urged to. He remembered having had the wild – and prophetic, he now realized – fear he might meet his brother and he ended in telemetrics, which was as far removed from genetic engineering as one could imagine. ... Or so one would have thought.

Then, through all the elaborate development of the Mercury Project, circumstance waited.

The time came, as it happened, when the Project appeared to be facing a dead end; and a suggestion had been made which saved the situation, and at the same time dragged Anthony into the dilemma his parents had prepared for him. And the best and most sardonic part of the whole thing was that it was Anthony who, in all innocence, made the suggestion.

2.

William Anti-Aut knew of the Mercury Project, but only in the way he knew of the long-drawn-out Stellar Probe that had been on its way long before he was born and would still be on its way after his death; and the way he knew of the Martian colony and of the continuing attempts to establish similar colonies on the asteroids.

Such things were on the distant periphery of his mind and of no real importance. No part of the space effort had ever swirled inward closer to the center of his interests, as far as he could remember, till the day when the printout included photographs of some of the men engaged in the Mercury Project.

William's attention was caught first by the fact that one of them had been identified as Anthony Smith. He remembered the odd name his brother had chosen, and he remembered the Anthony. Surely there could not be two Anthony Smiths.

He had then looked at the photograph itself and there was no mistaking the face. He looked in the mirror in a sudden whimsical gesture at checking the matter. No mistaking the face.

He felt amused, but uneasily so, for he did not fail to recognize the potentiality for embarrassment. Full blood brothers, to use the disgusting phrase. But what was there to do about it? How correct the fact that neither his father nor his mother had imagination?

He must have put the printout in his pocket, absently, when he was getting ready to leave for work, for he came across it at the lunch hour. He stared at it again. Anthony looked keen. It was quite a good reproduction – the printouts were of enormously good quality these days.

His lunch partner, Marco Whatever-his-name-was-that-week, said curiously, "What are you looking at, William?"

On impulse, William passed him the printout and said, "That's my brother." It was like grasping the nettle.

Marco studied it, frowning, and said, "Who? The man standing next to you?"

"No, the man who is me. I mean the man who looks like me. He's my brother."

There was a longer pause this time. Marco handed it back and said with a careful levelness to his voice, "Same-parents brother?"

"Yes."

"Father and mother both."

"Yes."

"Ridiculous!"

"I suppose so." William sighed. "Well, according to this, he's in telemetrics over in Texas and I'm doing work in autistics up here. So what difference does it make?"

William did not keep it in his mind and later that day he threw the printout away. He did not want his current bedmate to come across it. She had a ribald sense of humor that William was finding increasingly wearying. He was rather glad she was not in the mood for a child. He himself had had one a few years back anyway. That little brunette, Laura or Linda, one or the other name, had collaborated.

It was quite a time after that, at least a year, that the matter of Randall had come up. If William had given no further thought to his brother – and he hadn't – before that, he certainly had no time for it afterward.

Randall was sixteen when William first received word of him. He had lived a life that was increasingly seclusive and the Kentucky creche in which he was being brought up decided to cancel him and of course it was only some eight or ten days before cancellation that it occurred to anyone to report him to the New York Institute for the Science of Man. (The Homological Institute was its common name.)

William received the report along with reports of several others and there was nothing in the description of Randall that particularly attracted his notice. Still it was time for one of his tedious masstransport trips to the creches and there was one likely possibility in West Virginia. He went there – and was disappointed into swearing for the fiftieth time that he would thereafter make these visits by TV image – and then, having dragged himself there, thought he might as well take in the Kentucky creche before returning home.

He expected nothing.

Yet he hadn't studied Randall's gene pattern for more than ten minutes before he was calling the Institute for a computer calculation. Then he sat back and perspired slightly at the thought that only a last-minute impulse had brought him, and that without that impulse, Randall would have been quietly canceled in a week or less. To put it into the fine detail, a drug would have soaked painlessly through his skin and into his bloodstream and he would have sunk into a peaceful sleep that deepened gradually to death. The drug had a twenty-three-syllable official name, but William called it "nirvanamine," as did everyone else.

William said, "What is his full name, matron?"

The creche matron said, "Randall Nowan, scholar."

"No one!" said William explosively.

"Nowan." The matron spelled it. "He chose it last year."

"And it meant nothing to you? It is pronounced No one! It didn't occur to you to report this young man last year?"

"It didn't seem – " began the matron, flustered.

William waved her to silence. What was the use? How was she to know? There was nothing in the gene pattern to give warning by any of the usual textbook criteria. It was a subtle combination that William and his staff had worked out over a period of twenty years through experiments on autistic children – and a combination they had never actually seen in life.

So close to canceling!

Marco, who was the hardhead of the group, complained that the creches were too eager to abort before term and to cancel after term. He maintained that all gene patterns should be allowed to develop for purpose of initial screening and there should be no cancellation at all without consultation with a homologist.

"There aren't enough homologists," William said tranquilly.

"We can at least run all gene patterns through the computer," said Marco.

"To save anything we can get for our use?"

"For any homological use, here or elsewhere. We must study gene patterns in action if we're to understand ourselves properly, and it is the abnormal and monstrous patterns that give us most information. Our experiments on autism have taught us more about homology than the sum total existing on the day we began."

William, who still liked the roll of the phrase "the genetic physiology of man" rather than "homology," shook his head. "Just the same, we've got to play it carefully. However useful we can claim our experiments to be, we live on bare social permission, reluctantly given. We're playing with lives."

"Useless lives. Fit for canceling."

"A quick and pleasant canceling is one thing. Our experiments, usually long drawn out and sometimes unavoidably unpleasant, are another."

"We help them sometimes."

"And we don't help them sometimes."

It was a pointless argument, really, for there was no way of settling it. What it amounted to was that too few interesting abnormalities were available for homologists and there was no way of urging mankind to encourage a greater production. The trauma of the Catastrophe would never vanish in a dozen ways, including that one.

The hectic push toward space exploration could be traced back (and was, by some sociologists) to the knowledge of the fragility of the life skein on the planet, thanks to the Catastrophe.

Well, never mind.

There had never been anything like Randall Nowan. Not for William. The slow onset of autism characteristic of that totally rare gene pattern meant that more was known about Randall than about any equivalent patient before him. They even caught some last faint glimmers of his way of thought in the laboratory before he closed off altogether and shrank finally within the wall of his skin – unconcerned, unreachable.

Then they began the slow process whereby Randall, subjected for increasing lengths of time to artificial stimuli, yielded up the inner workings of his brain and gave clues thereby to the inner workings of all brains, those that were called normal as well as those like his own.

So vastly great was the data they were gathering that William began to feel his dream of reversing autism was more than merely a dream. He felt a warm gladness at having chosen the name Anti-Aut.

And it was at almost the height of the euphoria induced by the work on Randall that he received the call from Dallas and that the heavy pressure began – now, of all times – to abandon his work and take on a new problem.

Looking back on it later, he could never work out just what it was that finally led him to agree to visit Dallas. In the end, of course, he could see how fortunate it was – but what had persuaded him to do so? Could he, even at the start, have had a dim unrealized notion of what it might come to? Surely, impossible.

Was it the unrealized memory of that printout, that photograph of his brother? Surely, impossible.

But he let himself be argued into that visit and it was only when the micropile power unit changed the pitch of its soft hum and the agrav unit took over for the final descent that he remembered that photograph – or at least that it moved into the conscious part of his memory.

Anthony worked at Dallas and, William remembered now, at the Mercury Project. That was what the caption had referred to. He swallowed, as the soft jar told him the journey was over. This would be uncomfortable.

3.

Anthony was waiting on the roof reception area to greet the incoming expert. Not he by himself, of course. He was part of a sizable delegation – the size itself a rather grim indication of the desperation to which they had been reduced – and he was among the lower echelons. That he was there at all was only because it was he who had made the original suggestion.

He felt a slight, but continuing, uneasiness at the thought of that. He had put himself on the line. He had received considerable approval for it, but there had been the faint insistence always that it was his suggestion; and if it turned out to be a fiasco, every one of them would move out of the line of fire and leave him at point-zero.

There were occasions, later, when he brooded over the possibility that the dim memory of a brother in homology had suggested his thought. That might have been, but it didn't have to be. The suggestion was so sensibly inevitable, really, that surely he would have had the same thought if his brother had been something as innocuous as a fantasy writer, or if he had had no brother of his own.

The problem was the inner planets – The Moon and Mars were colonized. The larger asteroids and the satellites of Jupiter had been reached, and plans were in progress for a manned voyage to Titan, Saturn's large satellite, by way of an accelerating whirl about Jupiter. Yet even with plans in action for sending men on a seven-year round trip to the outer Solar System, there was still no chance of a manned approach to the inner planets, for fear of the Sun.

Venus itself was the less attractive of the two worlds within Earth's orbit. Mercury, on the other hand.

Anthony had not yet joined the team when Dmitri Large (he was quite short, actually) had given the talk that had moved the World Congress sufficiently to grant the appropriation that made the Mercury Project possible.

Anthony had listened to the tapes, and had heard Dmitri's presentation. Tradition was firm to the effect that it had been extemporaneous, and perhaps it was, but it was perfectly constructed and it held within it, in essence, every guideline followed by the Mercury Project since.

And the chief point made was that it would be wrong to wait until the technology had advanced to the point where a manned expedition through the rigors of Solar radiation could become feasible. Mercury was a unique environment that could teach much, and from Mercury's surface sustained observations could be made of the Sun that could not be made in any other way.

– Provided a man substitute – a robot, in short – could be placed on the planet.

A robot with the required physical characteristics could be built. Soft landings were as easy as kiss-my-hand. Yet once a robot landed, what did one do with him next?

He could make his observations and guide his actions on the basis of those observations, but the Project wanted his actions to be intricate and subtle, at least potentially, and they were not at all sure what observations he might make.

To prepare for all reasonable possibilities and to allow for all the intricacy desired, the robot would need to contain a computer (some at Dallas referred to it as a "brain," but Anthony scorned that verbal habit – perhaps because, he wondered later, the brain was his brother's field) sufficiently complex and versatile to fall into the same asteroid with a mammalian brain.

Yet nothing like that could be constructed and made portable enough to be carried to Mercury and landed there – or if carried and landed, to be mobile enough to be useful to the kind of robot they planned. Perhaps someday the positronic-path devices that the roboticists were playing with might make it possible, but that someday was not yet.

The alternative was to have the robot send back to Earth every observation it made the moment it was made, and a computer on Earth could then guide his every action on the basis of those observations. The robot's body, in short, was to be there, and his brain here.

Once that decision was reached, the key technicians were the telemetrists and it was then that Anthony joined the Project. He became one of those who labored to devise methods for receiving and returning impulses over distances of from 50 to 40 million miles, toward, and sometimes past, a Solar disk that could interfere with those impulses in a most ferocious manner.

He took to his job with passion and (he finally thought) with skill and success. It was he, more than anyone else, who had designed the three switching stations that had been hurled into permanent orbit about Mercury – the Mercury Orbiters. Each of them was capable of sending and receiving impulses from Mercury to Earth and from Earth to Mercury. Each was capable of resisting, more or less permanently, the radiation from the Sun, and more than that, each could filter out Solar interference.

Three equivalent Orbiters were placed at distance of a little over a million miles from Earth, reaching north and south of the plane of the Ecliptic so that they could receive the impulses from Mercury and relay them to Earth – or vice versa – even when Mercury was behind the Sun and inaccessible to direct reception from any station on Earth's surface.

Which left the robot itself; a marvelous specimen of the roboticists' and telemetrists' arts in combination. The most complex of ten successive models, it was capable, in a volume only a little over twice that of a man and five times his mass, of sensing and doing considerably more than a man – if it could be guided.

How complex a computer had to be to guide the robot made itself evident rapidly enough, however, as each response step had to be modified to allow for variations in possible perception. And as each response step itself enforced the certainty of greater complexity of possible variation in perceptions, the early steps had to be reinforced and made stronger. It built itself up endlessly, like a chess game, and the telemetrists began to use a computer to program the computer that designed the program for the computer that programmed the robot-controlling computer.

There was nothing but confusion. The robot was at a base in the desert spaces of Arizona and in itself was working well. The computer in Dallas could not, however, handle him well enough; not even under perfectly known Earth conditions. How then?

Anthony remembered the day when he had made the suggestion. It was on 7-4-553. He remembered it, for one thing, because he remembered thinking that day that 7-4 had been an important holiday in the Dallas region of the world among the pre-Cats half a millennium before – well, 553 years before, to be exact.

It had been at dinner, and a good dinner, too. There had been a careful adjustment of the ecology of the region and the Project personnel had high priority in collecting the food supplies that became available – so there was an unusual degree of choice on the menus, and Anthony had tried roast duck.

It was very good roast duck and it made him somewhat more expansive than usual. Everyone was in a rather self-expressive mood, in fact, and Ricardo said, "We'll never do it. Let's admit it. We'll never do it."

There was no telling how many had thought such a thing how many times before, but it was a rule that no one said so openly. Open pessimism might be the final push needed for appropriations to stop (they had been coming with greater difficulty each year for five years now) and if there were a chance, it would be gone.

Anthony, ordinarily not given to extraordinary optimism, but now reveling over his duck, said, "Why can't we do it? Tell me why, and I'll refute it."

It was a direct challenge and Ricardo's dark eyes narrowed at once. "You want me to tell you why?"

"I sure do." Ricardo swung his chair around, facing Anthony full. He said, "Come on, there's no mystery. Dmitri Large won't say so openly in any report, but you know and I know that to run Mercury Project properly, we'll need a computer as complex as a human brain whether it's on Mercury or here, and we can't build one. So where does that leave us except to play games with the World Congress and get money for make-work and possibly useful spin-offs?"

And Anthony placed a complacent smile on his face and said, "That's easy to refute. You've given us the answer yourself." (Was he playing games? Was it the warm feeling of duck in his stomach? The desire to tease Ricardo? ... Or did some unfelt thought of his brother touch him? There was no way, later, that he could tell.)

"What answer?" Ricardo rose. He was quite tall and unusually thin and he always wore his white coat unseamed. He folded his arms and seemed to be doing his best to tower over the seated Anthony like an unfolded meter rule. "What answer?"

"You say we need a computer as complex as a human brain. All right, then, we'll build one."

"The point, you idiot, is that we can't – "

"We can't. But there are others."

"What others?"

"People who work on brains, of course. We're just solid-state mechanics. We have no idea in what way a human brain is complex, or where, or to what extent. Why don't we get in a homologist and have him design a computer?" And with that Anthony took a huge helping of stuffing and savored it complacently. He could still remember, after all this time, the taste of the stuffing, though he couldn't remember in detail what had happened afterward.

It seemed to him that no one had taken it seriously. There was laughter and a general feeling that Anthony had wriggled out of a hole by clever sophistry so that the laughter was at Ricardo's expense. (Afterward, of course, everyone claimed to have taken the suggestion seriously.)

Ricardo blazed up, pointed a finger at Anthony, and said, "Write that up. I dare you to put that suggestion in writing." (At least, so Anthony's memory had it. Ricardo had, since then, stated his comment was an enthusiastic "Good idea! Why don't you write it up formally, Anthony?")

Either way, Anthony put it in writing.

Dmitri Large had taken to it. In private conference, he had slapped Anthony on the back and had said that he had been speculating in that direction himself – though he did not offer to take any credit for it on the record. (Just in case it turned out to be a fiasco, Anthony thought.)

Dmitri Large conducted the search for the appropriate homologist. It did not occur to Anthony that he ought to be interested. He knew neither homology nor homologists – except, of course, his brother, and he had not thought of him. Not consciously.

So Anthony was up there in the reception area, in a minor role, when the door of the aircraft opened and several men got out and came down and in the course of the handshakes that began going round, he found himself staring at his own face.

His cheeks burned and, with all his might, he wished himself a thousand miles away.

4.

More than ever, William wished that the memory of his brother had come earlier. It should have. ... Surely it should have.

But there had been the flattery of the request and the excitement that had begun to grow in him after a while. Perhaps he had deliberately avoided remembering.

To begin with, there had been the exhilaration of Dmitri Large coming to see him in his own proper presence. He had come from Dallas to New York by plane and that had been very titillating for William, whose secret vice it was to read thrillers. In the thrillers, men and women always traveled mass-wise when secrecy was desired. After all, electronic travel was public property – at least in the thrillers, where every radiation beam of whatever kind was invariably bugged.

William had said so in a kind of morbid half attempt at humor, but Dmitri hadn't seemed to be listening. He was staring at William's face and his thoughts seemed elsewhere. "I'm sorry," he said finally. "You remind me of someone."

(And yet that hadn't given it away to William. How was that possible? he had eventual occasion to wonder.)

Dmitri Large was a small plump man who seemed to be in a perpetual twinkle even when he declared himself worried or annoyed. He had a round and bulbous nose, pronounced cheeks, and softness everywhere. He emphasized his last name and said with a quickness that led William to suppose he said it often, "Size is not all the large there is, my friend."

In the talk that followed, William protested much. He knew nothing about computers. Nothing! He had not the faintest idea of how they worked or how they were programmed.

"No matter, no matter," Dmitri said, shoving the point aside with an expressive gesture of the hand. "We know the computers; we can set up the programs. You just tell us what it is a computer must be made to do so that it will work like a brain and not like a computer."

"I'm not sure I know enough about how a brain works to be able to tell you that, Dmitri," said William.

"You are the foremost homologist in the world," said Dmitri. "I have checked that out carefully." And that disposed of that.

William listened with gathering gloom. He supposed it was inevitable. Dip a person into one particular specialty deeply enough and long enough, and he would automatically begin to assume that specialists in all other fields were magicians, judging the depth of their wisdom by the breadth of his own ignorance. ... And as time went on, William learned a great deal more of the Mercury Project than it seemed to him at the time that he cared to.

He said at last, "Why use a computer at all, then? Why not have one of your own men, or relays of them, receive the material from the robot and send back instructions."

"Oh, oh, oh," said Dmitri, almost bouncing in his chair in his eagerness. "You see, you are not aware. Men are too slow to analyze quickly all the material the robot will send back – temperatures and gas pressures and cosmic-ray fluxes and Solar-wind intensities and chemical compositions and soil textures and easily three dozen more items – and then try to decide on the next step. A human being would merely guide the robot, and ineffectively; a computer would be the robot.

"And then, too," he went on, "men are too fast, also. It takes radiation of any kind anywhere from ten to twenty-two minutes to take the round trip between Mercury and Earth, depending on where each is in its orbit. Nothing can be done about that. You get an observation, you give an order, but much has happened between the time the observation is made and the response returns. Men can't adapt to the slowness of the speed of light, but a computer can take that into account. ... Come help us, William."

William said gloomily, "You are certainly welcome to consult me, for what good that might do you. My private TV beam is at your service."

"But it's not consultation I want. You must come with me."

"Mass-wise?" said William, shocked.

"Yes, of course. A project like this can't be carried out by sitting at opposite ends of a laser beam with a communications satellite in the middle. In the long run, it is too expensive, too inconvenient, and, of course, it lacks all privacy – "

It was like a thriller, William decided. "Come to Dallas," said Dmitri, "and let me show you what we have there. Let me show you the facilities. Talk to some of our computer men. Give them the benefit of your way of thought."

It was time, William thought, to be decisive. "Dmitri," he said, "I have work of my own here. Important work that I do not wish to leave. To do what you want me to do may take me away from my laboratory for months."

"Months!" said Dmitri, clearly taken aback. "My good William, it may well be years. But surely it will be your work."

"No, it will not. I know what my work is and guiding a robot on Mercury is not it."

"Why not? If you do it properly, you will learn more about the brain merely by trying to make a computer work like one, and you will come back here, finally, better equipped to do what you now consider your work. And while you're gone, will you have no associates to carry on? And can you not be in constant communication with them by laser beam and television? And can you not visit New York on occasion? Briefly."

William was moved. The thought of working on the brain from another direction did hit home. From that point on, he found himself looking for excuses to go – at least to visit – at least to see what it was all like. ... He could always return.

Then there followed Dmitri's visit to the ruins of Old New York, which he enjoyed with artless excitement (but then there was no more magnificent spectacle of the useless gigantism of the pre-Cats than Old New York). William began to wonder if the trip might not give him an opportunity to see some sights as well.

He even began to think that for some time he had been considering the possibility of finding a new bedmate, and it would be more convenient to find one in another geographical area where he would not stay permanently.

– Or was it that even then, when he knew nothing but the barest beginning of what was needed, there had already come to him, like the twinkle of a distant lightning flash, what might be done

So he eventually went to Dallas and stepped out on the roof and there was Dmitri again, beaming. Then, with eyes narrowing, the little man turned and said, "I knew – What a remarkable resemblance!"

William's eyes opened wide and there, visibly shrinking backward, was enough of his own face to make him certain at once that Anthony was standing before him.

He read very plainly in Anthony's face a longing to bury the relationship. All William needed to say was "How remarkable!" and let it go. The gene patterns of mankind were complex enough, after all, to allow resemblances of any reasonable degree even without kinship.

But of course William was a homologist and no one can work with the intricacies of the human brain without growing insensitive as to its details, so he said, "I'm sure this is Anthony, my brother."

Dmitri said, "Your brother?"

"My father," said William, "had two boys by the same woman – my mother. They were eccentric people."

He then stepped forward, hand outstretched, and Anthony had no choice but to take it. ... The incident was the topic of conversation, the only topic, for the next several days.

5.

It was small consolation to Anthony that William was contrite enough when he realized what he had done.

They sat together after dinner that night and William said, "My apologies. I thought that if we got the worst out at once that would end it. It doesn't seem to have done so. I've signed no papers, made no formal agreement. I will leave."

"What good would that do?" said Anthony ungraciously. "Everyone knows now. Two bodies and one face. It's enough to make one puke."

"If I leave – "

"You can't leave. This whole thing is my idea."

"To get me here?" William's heavy lids lifted as far as they might and his eyebrows climbed.

"No, of course not. To get a homologist here. How could I possibly know they would send you?"

"But if I leave – "

"No. The only thing we can do now is to lick the problem, if it can be done. Then – it won't matter." (Everything is forgiven those who succeed, he thought.)

"I don't know that I can – "

"We'll have to try. Dmitri will place it on us. It's too good a chance. You two are brothers," Anthony said, mimicking Dmitri's tenor voice, "and understand each other. Why not work together?" Then, in his own voice, angrily, "So we must. To begin with, what is it you do, William? I mean, more precisely than the word 'homology' can explain by itself."

William sighed. "Well, please accept my regrets. ... I work with autistic children."

"I'm afraid I don't know what that means."

"Without going into a long song and dance, I deal with children who do not reach out into the world, do not communicate with others, but who sink into themselves and exist behind a wall of skin, somewhat unreachably. I hope to be able to cure it someday."

"Is that why you call yourself Anti-Aut?"

"Yes, as a matter of fact."

Anthony laughed briefly, but he was not really amused.

A chill crept into William's manner. "It is an honest name."

"I'm sure it is," muttered Anthony hurriedly, and could bring himself to no more specific apology. With an effort, he restored the subject, "And are you making any progress?"

"Toward the cure? No, so far. Toward understanding, yes. And the more I understand – " William's voice grew warmer as he spoke and his eyes more distant. Anthony recognized it for what it was, the pleasure of speaking of what fills one's heart and mind to the exclusion of almost everything else. He felt it in himself often enough.

He listened as closely as he might to something he didn't really understand, for it was necessary to do so. He would expect William to listen to him.

How clearly he remembered it. He thought at the time he would not, but at the time, of course, he was not aware of what was happening. Thinking back, in the glare of hindsight, he found himself remembering whole sentences, virtually word for word.

"So it seemed to us," William said, "that the autistic child was not failing to receive the impressions, or even failing to interpret them in quite a sophisticated manner. He was, rather, disapproving them and rejecting them, without any loss of the potentiality of full communication if some impression could be found which he approved of."

"Ah," said Anthony, making just enough of a sound to indicate that he was listening.

"Nor can you persuade him out of his autism in any ordinary way, for he disapproves of you just as much as he disapproves of the rest of the world. But if you place him in conscious arrest – "

"In what?"

"It is a technique we have in which, in effect, the brain is divorced from the body and can perform its functions without reference to the body. It is a rather sophisticated technique devised in our own laboratory; actually – " He paused.

"By yourself?" asked Anthony gently. "Actually, yes," said William, reddening slightly, but clearly pleased. "In conscious arrest, we can supply the body with designed fantasies and observe the brain under differential electroencephalography. We can at once learn more about the autistic individual; what kind of sense impressions he most wants; and we learn more about the brain generally."

"Ah," said Anthony, and this time it was a real ah. "And all this you have learned about brains – can you not adapt it to the workings of a computer?"

"No," said William. "Not a chance. I told that to Dmitri. I know nothing about computers and not enough about brains."

"If I teach you about computers and tell you in detail what we need, what then?"

"It won't do. It – "

"Brother," Anthony said, and he tried to make it an impressive word. "You owe me something. Please make an honest attempt to give our problem some thought. Whatever you know about the brain – please adapt it to our computers."

William shifted uneasily, and said, "I understand your position. I will try. I will honestly try."

6.

William had tried, and as Anthony had predicted, the two had been left to work together. At first they encountered others now and then and William had tried to use the shock value of the announcement that they were brothers since there was no use in denial. Eventually that stopped, however, and there came to be a purposeful non-interference. When William approached Anthony, or Anthony approached William, anyone else who might be present faded silently into the walls.

They even grew used to each other after a fashion and sometimes spoke to each other almost as though there were no resemblance between them at all and no childish memories in common.

Anthony made the computer requirements plain in reasonably non-technical language and William, after long thought, explained how it seemed to him a computer might do the work, more or less, of a brain.

Anthony said, "Would that be possible?"

"I don't know," said William. "I am not eager to try. It may not work. But it may."

"We'd have to talk to Dmitri Large."

"Let's talk it over ourselves first and see what we've got. We can go to him with as reasonable a proposition as we can put together. Or else, not go to him."

Anthony hesitated, "We both go to him?" William said delicately, "You be my spokesman. There is no reason that we need be seen together."

"Thank you, William. If anything comes of this, you will get full credit from me."

William said, "I have no worries about that. If there is anything to this, I will be the only one who can make it work, I suppose."

They thrashed it out through four or five meetings and if Anthony hadn't been kin and if there hadn't been that sticky, emotional situation between them, William would have been uncomplicatedly proud of the younger-brother – for his quick understanding of an alien field.

There were then long conferences with Dmitri Large. There were, in fact, conferences with everyone. Anthony saw them through endless days, and then they came to see William separately. And eventually, through an agonizing pregnancy, what came to be called the Mercury Computer was authorized.

William then returned to New York with some relief. He did not plan to stay in New York (would he have thought that possible two months earlier?) but there was much to do at the Homological Institute.

More conferences were necessary, of course, to explain to his own laboratory group what was happening and why he had to take leave and how they were to continue their own projects without him. Then there was a much more elaborate arrival at Dallas with the essential equipment and with two young aides for what would have to be an open-ended stay.

Nor did William even look back, figuratively speaking. His own laboratory and its needs faded from his thoughts. He was now thoroughly committed to his new task.

7.

It was the worst period for Anthony. The relief during William's absence had not penetrated deep and there began the nervous agony of wondering whether perhaps, hope against hope, he might not return. Might he not choose to send a deputy, someone else, anyone else? Anyone with a different face so that Anthony need not feel the half of a two-backed four-legged monster?

But it was William. Anthony had watched the freight plane come silently through the air, had watched it unload from a distance. But even from that distance he eventually saw William.

That was that. Anthony left. He went to see Dmitri that afternoon. "It's not necessary, Dmitri, for me to stay, surely. We've worked out the details and someone else can take over."

"No, no," said Dmitri. "The idea was yours in the first place. You must see it through. There is no point in needlessly dividing the credit."

Anthony thought: No one else will take the risk. There's still the chance of fiasco. I might have known.

He had known, but he said stolidly, "You understand I cannot work with William."

"But why not?" Dmitri pretended surprise. "You have been doing so well together."

"I have been straining my guts over it, Dmitri, and they won't take any more. Don't you suppose I know how it looks?"

"My good fellow! You make too much of it. Sure the men stare. They are human, after all. But they'll get used to it. I'm used to it."

You are not, you fat liar, Anthony thought. He said, "I'm not used to it."

"You're not looking at it properly. Your parents were peculiar – but after all, what they did wasn't illegal, only peculiar, only peculiar. It's not your fault, or William's. Neither of you is to blame."

"We carry the mark," said Anthony, making a quick curving gesture of his hand to his face.

"It's not the mark you think. I see differences. You are distinctly younger in appearance. Your hair is wavier. It's only at first glance that there is a similarity. Come, Anthony, there will be all the time you want, all the help you need, all the equipment you can use. I'm sure it will work marvelously. Think of the satisfaction – "

Anthony weakened, of course, and agreed at least to help William set up the equipment. William; too, seemed sure it would work marvelously. Not as frenetically as Dmitri did, but with a kind of calmness.

"It's only a matter of the proper connections," he said, "though I must admit that that's quite a huge 'only.' Your end of it will be to arrange sensory impressions on an independent screen so that we can exert – well, I can't say manual control, can I? – so that we can exert intellectual control to override, if necessary."

"That can be done," said Anthony. "Then let's get going. ... Look, I'll need a week at least to arrange the connections and make sure of the instructions – "

"Programming," said Anthony. "Well, this is your place, so I'll use your terminology. My assistants and I will program the Mercury Computer, but not in your fashion."

"I should hope not. We would want a homologist to set up a much more subtle program than anything a mere telemetrist could do." He did not try to hide the self-hating irony in his words.

William let the tone go and accepted the words. He said, "We'll begin simply. We'll have the robot walk."

8.

A week later, the robot walked in Arizona, a thousand miles away. He walked stiffly, and sometimes he fell down, and sometimes he clanked his ankle against an obstruction, and sometimes he whirled on one foot and went off in a surprising new direction.

"He's a baby, learning to walk," said William. Dmitri came occasionally, to learn of progress. "That's remarkable," he would say.

Anthony didn't think so. Weeks passed, then months. The robot had progressively done more and more, as the Mercury Computer had been placed, progressively, under a more and more complex programming. (William had a tendency to refer to the Mercury Computer as a brain, but Anthony wouldn't allow it.) And all that happened wasn't good enough.

"It's not good enough, William," he said finally. He had not slept the night before.

"Isn't that strange?" said William coolly. "I was going to say that I thought we had it about beaten."

Anthony held himself together with difficulty. The strain of working with William and of watching the robot fumble was more than he could bear. "I'm going to resign, William. The whole job. I'm sorry. ... It's not you."

"But it is I, Anthony."

"It isn't all you, William. It's failure. We won't make it. You see how clumsily the robot handles himself, even though he's on Earth, only a thousand miles away, with the signal round trip only a tiny fraction of a second in time. On Mercury, there will be minutes of delay, minutes for which the Mercury Computer will have to allow. It's madness to think it will work."

William said, "Don't resign, Anthony. You can't resign now. I suggest we have the robot sent to Mercury. I'm convinced he's ready."

Anthony laughed loudly and insultingly. "You're crazy, William."

"I'm not. You seem to think it will be harder on Mercury, but it won't be. It's harder on Earth. This robot is designed for one-third Earth-normal gravity, and he's working in Arizona at full gravity. He's designed for 400ø C, and he's got 300ø C. He's designed for vacuum and he's working in an atmospheric soup."

"That robot can take the difference."

"The metal structure can, I suppose, but what about the Computer right here? It doesn't work well with a robot that isn't in the environment he's designed for. ... Look, Anthony, if you want a computer that is as complex as a brain, you have to allow for idiosyncrasies. ... Come, let's make a deal. If you will push, with me, to have the robot sent to Mercury, that will take six months, and I will take a sabbatical for that period. You will be rid of me."

"Who'll take care of the Mercury Computer?"

"By now you understand how it works, and I'll have my two men here to help you."

Anthony shook his head defiantly. "I can't take the responsibility for the Computer, and I won't take the responsibility for suggesting that the robot be sent to Mercury. It won't work."

"I'm sure it will."

"You can't be sure. And the responsibility is mine. I'm the one who'll bear the blame. It will be nothing to you."

Anthony later remembered this as a crucial moment. William might have let it go. Anthony would have resigned. All would have been lost.

But William said, "Nothing to me? Look, Dad had this thing about Mom. All right. I'm sorry, too. I'm as sorry as anyone can be, but it's done, and there's something funny that has resulted. When I speak of Dad, I mean your Dad, too, and there's lots of pairs of people who can say that: two brothers, two sisters, a brother and sister. And then when I say Mom, I mean your Mom, and there are lots of pairs who can say that, too. But I don't know any other pair, nor have I heard of any other pair, who can share both Dad and Mom."

"I know that," said Anthony grimly. "Yes, but look at it from my standpoint," said William hurriedly. "I'm a homologist. I work with gene patterns. Have you ever thought of our gene patterns? We share both parents, which means that our gene patterns are closer together than any other pair on this planet. Our very faces show it."

"I know that, too."

"So that if this project were to work, and if you were to gain glory from it, it would be your gene pattern that would have been proven highly useful to mankind – and that would mean very much my gene pattern as well. ... Don't you see, Anthony? I share your parents, your face, your gene pattern, and therefore either your glory or your disgrace. It is mine almost as much as yours, and if any credit or blame adheres to me, it is yours almost as much as mine, too. I've got to be interested in your success. I've a motive for that which no one else on Earth has – a purely selfish one, one so selfish you can be sure it's there. I'm on your side, Anthony, because you're very nearly me!"

They looked at each other for a long time, and for the first time, Anthony did so without noticing the face he shared.

William said, "So let us ask that the robot be sent to Mercury."

And Anthony gave in. And after Dmitri had approved the request – he had been waiting to, after all – Anthony spent much of the day in deep thought.

Then he sought out William and said, "Listen!"

There was a long pause which William did not break. Anthony said again, "Listen!" William waited patiently.

Anthony said, "There's really no need for you to leave. I'm sure you wouldn't like to have the Mercury Computer tended by anyone but yourself."

William said, "You mean you intend to leave?" Anthony said, "No, I'll stay, too."

William said, "We needn't see much of each other."

All of this had been, for Anthony, like speaking with a pair of hands clenched about his windpipe. The pressure seemed to tighten now, but he managed the hardest statement of all.

"We don't have to avoid each other. We don't have to."

William smiled rather uncertainly. Anthony didn't smile at all; he left quickly.

9.

William looked up from his book. It was at least a month since he had ceased being vaguely surprised at having Anthony enter.

He said, "Anything wrong?"

"Who can say? They're coming in for the soft landing. Is the Mercury Computer in action?"

William knew Anthony knew the Computer status perfectly, but he said, "By tomorrow morning, Anthony."

"And there are no problems?"

"None at all."

"Then we have to wait for the soft landing."

"Yes."

Anthony said, "Something will go wrong."

"Rocketry is surely an old hand at this. Nothing will go wrong."

"So much work wasted."

"It's not wasted yet. It won't be."

Anthony said, "Maybe you're right." Hands deep in his pockets, he drifted away, stopping at the door just before touching contact. "Thanks!"

"For what, Anthony?"

"For being – comforting."

William smiled wryly and was relieved his emotions didn't show.

10.

Virtually the entire body of personnel of the Mercury Project was on hand for the crucial moment. Anthony, who had no tasks to perform, remained well to the rear, his eyes on the monitors. The robot had been activated and there were visual messages being returned.

At least they came out as the equivalent of visual – and they showed as yet nothing but a dim glow of light which was, presumably, Mercury's surface.

Shadows flitted across the screen, probably irregularities on that surface. Anthony couldn't tell by eye alone, but those at the controls, who were analyzing the data by methods more subtle than could be disposed of by unaided eye, seemed calm. None of the little red lights that might have betokened emergency were lighting. Anthony was watching the key observers rather than the screen.

He should be down with William and the others at the Computer. It was going to be thrown in only when the soft landing was made. He should be. He couldn't be.

The shadows flitted across the screen more rapidly. The robot was descending – too quickly? Surely, too quickly!

There was a last blur and a steadiness, a shift of focus in which the blur grew darker, then fainter. A sound was heard and there were perceptible seconds before Anthony realized what it was the sound was saying – "Soft landing achieved! Soft landing achieved!"

Then a murmur arose and became an excited hum of self-congratulation until one more change took place on the screen and the sound of human words and laughter was stopped as though there had been a smash collision against a wall of silence.

For the screen changed; changed and grew sharp. In the brilliant, brilliant sunlight, blazing through the carefully filtered screen, they could now see a boulder clear, burning white on one side, ink-on-ink on the other. It shifted right, then back to left, as though a pair of eyes were looking left, then right. A metal hand appeared on the screen as though the eyes were looking at part of itself.

It was Anthony's voice that cried out at last, "The Computer's been thrown in."

He heard the words as though someone else had shouted them and he raced out and down the stairs and through a Corridor, leaving the babble of voices to rise behind him.

"William," he cried as he burst into the Computer room, "it's perfect, it's – "

But William's hand was upraised. "Shh. Please. I don't want any violent sensations entering except those from the robot."

"You mean we can be heard?" whispered Anthony.

"Maybe not, but I don't know." There was another screen, a smaller one, in the room with the Mercury Computer. The scene on it was different, and changing; the robot was moving.

William said, "The robot is feeling its way. Those steps have got to be clumsy. There's a seven-minute delay between stimulus and response and that has to be allowed for."

"But already he's walking more surely than he ever did in Arizona. Don't you think so, William? Don't you think so?" Anthony was gripping William's shoulder, shaking it, eyes never leaving the screen.

William said, "I'm sure of it, Anthony."

The Sun burned down in a warm contrasting world of white and black, of white Sun against black sky and white rolling ground mottled with black shadow. The bright sweet smell of the Sun on every exposed square centimeter of metal contrasting with the creeping death-of-aroma on the other side.

He lifted his hand and stared at it, counting the fingers. Hot-hot-hot-turning, putting each finger, one by one, into the shadow of the others and the hot slowly dying in a change in tactility that made him feel the clean, comfortable vacuum.

Yet not entirely vacuum. He straightened and lifted both arms over his head, stretching them out, and the sensitive spots on either wrist felt the vapors – the thin, faint touch of tin and lead rolling through the cloy of mercury.

The thicker taste rose from his feet; the silicates of each variety, marked by the clear separate-and-together touch and tang of each metal ion. He moved one foot slowly through the crunchy, caked dust, and felt the changes like a soft, not quite random symphony.

And over all the Sun. He looked up at it, large and fat and bright and hot, and heard its joy. He watched the slow rise of prominences around its rim and listened to the crackling sound of each; and to the other happy noises over the broad face. When he dimmed the background light, the red of the rising wisps of hydrogen showed in bursts of mellow contralto, and the deep bass of the spots amid the muted whistling of the wispy, moving faculae, and the occasional thin keening of a flare, the ping-pong ticking of gamma rays and cosmic particles, and over all in every direction the soft, fainting, and ever-renewed sigh of the Sun's substance rising and retreating forever in a cosmic wind which reached out and bathed him in glory.

He jumped, and rose slowly in the air with a freedom he had never felt, and jumped again when he landed, and ran, and jumped, and ran again, with a body that responded perfectly to this glorious world, this paradise in which he found himself.

A stranger so long and so lost – in paradise at last.

William said, "It's all right."

"But what's he doing?" cried out Anthony.

"It's all right. The programming is working. He has tested his senses. He has been making the various visual observations. He has dimmed the Sun and studied it. He has tested for atmosphere and for the chemical nature of the soil. It all works."

"But why is he running?"

"I rather think that's his own idea, Anthony. If you want to program a computer as complicated as a brain, you've got to expect it to have ideas of its own."

"Running? Jumping?" Anthony turned an anxious face to William. "He'll hurt himself. You can handle the Computer. Override. Make him stop."

And William said sharply, "No. I won't. I'll take the chance of his hurting himself. Don't you understand? He's happy. He was on Earth, a world he was never equipped to handle. Now he's on Mercury with a body perfectly adapted to its environment, as perfectly adapted as a hundred specialized scientists could make it be. It's paradise for him; let him enjoy it."

"Enjoy? He's a robot."

"I'm not talking about the robot. I'm talking about the brain – the brain – that's living here."

The Mercury Computer, enclosed in glass, carefully and delicately wired, its integrity most subtly preserved, breathed and lived.

"It's Randall who's in paradise," said William. "He's found the world for whose sake he autistically fled this one. He has a world his new body fits perfectly in exchange for the world his old body did not fit at all."

Anthony watched the screen in wonder. "He seems to be quieting."

"Of course," said William, "and he'll do his job all the better for his joy."

Anthony smiled and said, "We've done it, then, you and I? Shall we join the rest and let them fawn on us, William?"

William said, "Together?"

And Anthony linked arms. "Together, brother!"

After You Have Read the Story ...

Randall is more than a *teleoperator,* isn't he? Asimov struck a chord that resonates beyond the use of robots for prosthetic legs and arms; Randall finally gets a body that matches his brain. Even when Asimov wrote the story in 1970, autism was already emerging and becoming part of the social consciousness. The story reflects the very narrow and pessimistic view of autism and often unintentionally cruel treatment found in society at that time, but the message of freeing a child imprisoned by circumstance still resonates.

What Asimov, or anyone back then, couldn't foresee was the role of robotics in autism research and therapy. Researchers such as Brian Scazellati and Aude Billiard found that many children on the autism spectrum could relate to simple, mechanical robot heads. Those children are often overwhelmed by the hundreds of muscles in people's faces and resultant numerous twitches and by the constant saccading of their own eyes. They don't have built-in neurological filters to help them focus on the most important facial cues and maintain eye contact. Robot heads give children on one portion of the autism spectrum a less demanding target that does not distract them as they are taught to read human faces.

On a more mundane level, Asimov's story provides a starting point for introducing the history and components of telesystems, the engineering guidelines for choosing a telesystem as a robotic solution (described in chapter 5 of *Introduction to AI Robotics*[4]), and other concepts such as behavior-based robotics and situated agency.

History of Telesystems

In terms of artificial intelligence, telesystems were originally treated as a degenerative case of full autonomy, sometimes called "level 0," which was not a compliment. Full autonomy would be having a robot that could be assigned to perform an entire task by itself; this is sometimes called *taskable agency* because the robot is a distinct entity, or "agent" in AI parlance, that can be delegated tasks.

Throughout the '70s, '80s, and '90s, researchers thought that telepresence would solve the challenges of *remote presence*—if the operator could somehow be immersed in the remote environment, the human operator's brain wouldn't notice any difference and thus would not incur any additional cognitive workload. In practice, however, telepresence robots are fairly rare as robots do not seamlessly match the movements of a person, the sensors are incomplete (we say the operator is "sensor impoverished"), the sensor output is not the same stimulus as what we experience, and there remains a significant latency in sensing and executing an action.

Regardless of the excitement over telepresence and immersive interfaces, telesystems were generally viewed as uncool until about 2010. If a scientist specializing in artificial intelligence said they were working on telesystem research or the related area of human supervisory control, they would likely get sad looks from other researchers and perhaps wind up eating lunch alone. But in the late 2000s, researchers recognized that taskable agency is not desirable for all missions and that for many missions, such as search and rescue, the whole goal is for a responder to sense and act remotely in real time rather than wait until a robot returns home and uploads the captured data for the responder to review. These remote presence missions introduce the challenges of a human and robot working cooperatively together.

In the 2010s, teleoperation went from being tolerated to fashionable as the telecommuting market opened up. Even now, telecommuting through a commercial robot shows that telesystems are not just a straightforward application of robots. A telecommunicating robot is a multimedia interface on wheels, but it is not the same rich sensory experience as being in the workplace. The operator has to mentally translate intentions such as walking toward another person into joystick commands and doesn't have the same field of view and general sense of situatedness as being there. Now telesystems are no longer viewed as a degenerative state of autonomy but instead as one of many styles of shared human-robot intelligence.

Components of a Telesystem

The story actually does a fairly decent job of describing the seven components of a teleystem. In telesystem parlance, the local unit, or operator control unit, houses two of the seven basic components:

1) some type of *display* of what the remote unit is doing and its status and
2) some method of controlling the remote unit called the *local control device*.

Presumably the local control device is the user interface by which a teleoperator controls the robot; in cognitive engineering the human-robot interaction required to make the local control device work easily and reliably remains an open research question. The remote unit houses four of the seven components:

3) the *remote control device* or whatever onboard intelligence is needed for the unit to operate,
4) the device's *sensors*,
5) the device's appendages and mechanisms for moving and picking up things called *effectors*, and
6) the device's *power*.

The remote unit was originally called the *telefactor*, with the telesystem being a teleoperator plus a telefactor, but that terminology is less common now. The final component is

7) *communication* between the local and remote units

and it is this component that serves ostensibly as the motivation for a novel local control device and brings the two brothers together to find a solution.

In the story, the local control device for the robot is in Dallas, Texas, not Houston, even though NASA Johnson Manned Space Center has been based in Houston since 1961, a decade before the story was written. The remote unit is the robot and it is located on Mercury. Asimov focuses on a novel local control device to overcome the problems with the communication link. The link has sufficient bandwidth due to three repeaters but still has a seven minute delay, which is too hard for human teleoperators to adapt to.

The time lag or latency is a pernicious problem in teleoperation and is even worse than just the signal transit time cited in the story. The teleoperation delay heuristic states that the time it takes to do a task with traditional teleoperation grows linearly with the transmission delay. A teleoperation task that took 1 minute for a teleoperator to guide a remote to do on Earth might take 2.5 minutes to do on the moon, and 140 minutes on Mars.[5] Mars planetary explorers all have onboard intelligence to move and react to obstacles and situations without intervention from the teleoperator.

The robot reads as if it is a humanoid but larger than a person with five times the mass, plus implied different sensors. It's not clear why a robot would be humanoid but that could make the mapping from the teleoperator's (Randall's) frame of reference to the telefactor easier. The power and the onboard intelligence needed for the robot to translate commands from the local control device and do other "hidden" functions such as maintain internal health diagnostics are not discussed but would be included in a real robot.

Guidelines for Choosing a Telesystem

The story sets up teleoperation as a work around because the project managers did not have a robot with the brains to work as a peer. But exploring Mecury appears to be a time-critical mission from which they do want the sort of observations that the managers believe they can only get from a human. This suggests that teleoperation, or telepresence, was the real desire. Despite being written more than forty years ago, the story mirrors the ongoing debate between manned and unmanned space exploration.

Charles Wampler reports the following guidelines for the choice of a telesystem for a particular project:[6]

1. The tasks are unstructured and not repetitive.
2. The task workspace cannot be engineered to permit the use of industrial manipulators.
3. Key portions of the task intermittently require dexterous manipulation, especially hand-eye coordination.
4. Key portions of the task require object recognition, situational awareness, or other advanced perception.
5. The needs of the display technology do not exceed the limitations of the communication link (bandwidth, time delays).
6. The availability of trained personnel is not an issue.

The choice of a telesystem in the story seems consistent with good engineering as the Mercury Project scores a perfect 6 out of 6 for suitability. The types of tasks that the managers want Randall to do are unstructured and not repetitive. Certainly Mercury cannot be engineered to simplify robotics and the environment on the planet is largely unmodeled, which is the point of exploration. Randall's tasks will certainly require dexterous manipulation, especially hand-eye coordination, and key portions of the task require object recognition, situational awareness, or other advanced perception. Asimov even says, "the Project wanted his actions to be intricate and subtle, at least potentially, and they were not at all sure what observations he might make." The needs of the display technology do not exceed the limitations of the communication link that has been engineered to have sufficient bandwidth. The story is technically weak here as it does not explain how autism overcomes the latency issue, implying only that the plasticity of Randall's brain will allow him to learn. Ultimately, the availability of trained personnel is not an issue because the heart, and literal brains, of the story is about a specific person whose entire existence was a match for the project.

Behavior-Based Robotics and Situated Agency

"Stranger in Paradise" touches on biological inspirations for robots, which are the subject of a formal subdiscipline called *behavior-based robotics* (covered in my textbook *Introduction to AI Robotics*[7]). Unable to engineer a solution on their own, the project managers turn to biology for inspiration. A similar phenomenon happened in real life in the mid-1980s. After an initial impressive success with the first mobile robot Shakey in 1967,[8] artificial intelligence for robotics made very little progress. Then in the mid-1980s a group of researchers led by Rod Brooks, Ron Arkin, and Dave Payton began to explore and replicate animal intelligence, leading to major breakthroughs.

The story also indirectly addresses *situated agency*, the idea that a robot is physically situated in the world and thus can, and should be allowed to, react to that world. Note that Randall is able to effectively control the robot only when it is in the ecology for which it was designed.

Reality Score: A

Asimov got the technology of telesystems right. Regardless of whether you like the story or are offended by its primitive view of autism, it should be clear that telepresence is not a degenerative case of intelligence but offers the opportunity to engage a person's unique intelligence and enable them to sense and act at a distance.

2 Behavior-Based Robotics: "Runaround"

Unlike "Stranger in Paradise," Asimov's "Runaround" is in the Three Laws of Robotics canon. It was published in 1942, making it the first of his stories that made use of the Three Laws.[1] The media often treats the Three Laws as the ultimate expression of robot ethics, but Asimov deliberately constructed the laws to sound reasonable while actually being ambiguous, thus creating conflicts that in turn generate storylines. The laws are less about ethical decisions and more about basic engineering safety guidelines to prevent accidents. The contribution of "Runaround" to the understanding of robotics is not, however, in its application of the Three Laws but in its exploration of the concepts of *behavior-based robotics* and *potential field methodologies*, the latter of which is a category of mechanisms for expressing behaviors.

From a technical standpoint, "Runaround" describes the Three Laws as if they are producing competing forces acting on the robot, Speedy. One way to read the story is that each law is an independent program or subroutine that produces a unique *behavior*. Each law or behavior literally produces a vector as if from gravitational or magnetic fields. These types of vector fields are part of a general class called *potential fields*. The interaction between the laws, and the vectors they produce, describe the direction that Speedy the robot takes as it drives around Mercury. Potential field methodologies are common in behavior-based robotics, adding a bit of realism to the story. The technological MacGuffin in this story is that vectors from potential fields can cancel each other out (which is true in real life), leading to unintended consequences.

As You Read the Story ...

As you read, it will be helpful to think of the robot from two perspectives. One is that of an animal ethnologist who is trying to break down the robot's action into the equivalent animal behaviors—for example, "the urge to do X," "the urge to do Y," and so on. The other is the perspective of a physicist who is modeling the output of each behavior as a vector that translates the urge into a movement. This is the same as computing the trajectory of a baseball, with one vector representing the ball's straight path and another vector representing the pull of gravity. The vectors sum to a single resultant vector that combines all the forces, or in the case of animals, all their behaviors.

Recall that vectors represent a force as an arrow and that different forces, or vectors, can be combined with vector addition. The direction of the arrow is the direction, or orientation, of the force and the length of the arrow is its magnitude. In robotics, just as in animal behavior, a vector would represent the direction in which the robot is heading and the magnitude or velocity with which it is moving. Vectors naturally translate into steering and speed commands.

Vector addition allows vectors to be added, and it computes the resultant vector. Figure 2.1 shows a robot "sensing" an attractive force from a goal, Vector 1, that makes the robot want to turn 45 degrees and move toward the goal with a velocity of 1. But the robot also senses an obstacle to avoid that produces a repulsive force, Vector 2, and it makes the robot want to turn 198 degrees and move at a velocity of 1 to stay away. A graphical mechanism for summing vectors is shown; there is also a more accurate mathematical formula that can be found in any physics textbook. The graphical mechanism is more intuitive and so is used here. The process is simple: the tail of Vector 2 is placed at the head of Vector 1, and then a line is drawn from the tail of Vector 1 to the head of Vector 2, which produces a resultant vector with the correct magnitude and orientation. The resultant force with which the robot actually moves is approximately 120 degrees with a velocity of 0.6. When it moves and senses the goal and obstacle in the next update cycle, the robot will still feel the two forces but the directions will now be different. The new result will send the robot closer to the goal but still keep it away from the obstacle, allowing it to move slowly until the robot finally reaches the goal.

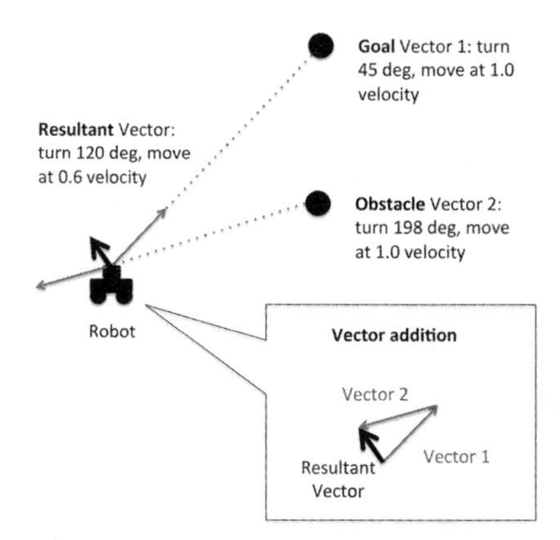

Figure 2.1
Example of graphical vector addition.

As you read, see if you can draw out the vectors that might explain Speedy's overall behavior. The "After You Have Read the Story" section will discuss behaviors in more depth, the five primitive potential fields used to construct more sophisticated behaviors, and the local minima problem Speedy encountered that is often cited, incorrectly, as a reason not to use potential fields.

"Runaround" by Isaac Asimov, 1942

It was one of Gregory Powell's favorite platitudes that nothing was to be gained from excitement, so when Mike Donovan came leaping down the stairs toward him, red hair matted with perspiration, Powell frowned.

"What's wrong?" he said. "Break a fingernail?"

"Yaaaah," snarled Donovan, feverishly. "What have you been doing in the sublevels all day?" He took a deep breath and blurted out, "Speedy never returned."

Powell's eyes widened momentarily and he stopped on the stairs; then he recovered and resumed his upward steps. He didn't speak until he reached the head of the flight, and then:

"You sent him after the selenium?"

"Yes."

"And how long has he been out?"

"Five hours now."

Silence! This was a devil of a situation. Here they were, on Mercury exactly twelve hours – and already up to the eyebrows in the worst sort of trouble. Mercury had long been the jinx world of the System, but this was drawing it rather strong – even for a jinx.

Powell said, "Start at the beginning, and let's get this straight."

They were in the radio room now – with its already subtly antiquated equipment, untouched for the ten years previous to their arrival. Even ten years, technologically speaking, meant so much. Compare Speedy with the type of robot they must have had back in 2005. But then, advances in robotics these days were tremendous. Powell touched a still gleaming metal surface gingerly. The air of disuse that touched everything about the room – and the entire Station – was infinitely depressing.

Donovan must have felt it. He began: "I tried to locate him by radio, but it was no go. Radio isn't any good on the Mercury Sunside – not past two miles, anyway. That's one of the reasons the First Expedition failed. And we can't put up the ultrawave equipment for weeks yet – "

"Skip all that. What did you get?"

"I located the unorganized body signal in the short wave. It was no good for anything except his position. I kept track of him that way for two hours and plotted the results on the map."

There was a yellowed square of parchment in his hip pocket – a relic of the unsuccessful First Expedition – and he slapped it down on the desk with vicious force, spreading it flat with the palm of his hand. Powell, hands clasped across his chest, watched it at long range.

Donovan's pencil pointed nervously. "The red cross is the selenium pool. You marked it yourself."

"Which one is it?" interrupted Powell. "There were three that MacDougal located for us before he left."

"I sent Speedy to the nearest, naturally; seventeen miles away. But what difference does that make?" There was tension in his voice. "There are the penciled dots that mark Speedy's position."

And for the first time Powell's artificial aplomb was shaken and his hands shot forward for the map.

"Are you serious? This is impossible."

"There it is," growled Donovan.

The little dots that marked the position formed a rough circle about the red cross of the selenium pool. And Powell's fingers went to his brown mustache, the unfailing signal of anxiety.

Donovan added: "In the two hours I checked on him, he circled that damned pool four times. It seems likely to me that he'll keep that up forever. Do you realize the position we're in?"

Powell looked up shortly, and said nothing. Oh, yes, he realized the position they were in. It worked itself out as simply as a syllogism. The photocell banks that alone stood between the full power of Mercury's monstrous sun and themselves were shot to hell.

The only thing that could save them was selenium. The only thing that could get the selenium was Speedy. If Speedy didn't come back, no selenium. No selenium, no photocell banks. No photo-banks – well, death by slow broiling is one of the more unpleasant ways of being done in.

Donovan rubbed his red mop of hair savagely and expressed himself with bitterness. "We'll be the laughingstock of the System, Greg. How can everything have gone so wrong so soon? The great team of Powell and Donovan is sent out to Mercury to report on the advisability of reopening the Sunside

Mining Station with modern techniques and robots and we ruin everything the first day. A purely routine job, too. We'll never live it down."

"We won't have to, perhaps," replied Powell, quietly. "If we don't do something quickly, living anything down – or even just plain living – will be out of the question."

"Don't be stupid! If you feel funny about it, Greg, I don't. It was criminal, sending us out here with only one robot. And it was your bright idea that we could handle the photocell banks ourselves."

"Now you're being unfair. It was a mutual decision and you know it. All we needed was a kilogram of selenium, a Stillhead Dielectrode Plate and about three hours' time and there are pools of pure selenium all over Sunside. MacDougal's spectroreflector spotted three for us in five minutes, didn't it? What the devil! We couldn't have waited for next conjunction."

"Well, what are we going to do? Powell, you've got an idea. I know you have, or you wouldn't be so calm. You're no more a hero than I am. Go on, spill it!"

"We can't go after Speedy ourselves, Mike – not on the Sunside. Even the new insosuits aren't good for more than twenty minutes in direct sunlight. But you know the old saying, 'Set a robot to catch a robot.' Look, Mike, maybe things aren't so bad. We've got six robots down in the sublevels, that we may be able to use, if they work. If they work."

There was a glint of sudden hope in Donovan's eyes. "You mean six robots from the First Expedition. Are you sure? They may be subrobotic machines. Ten years is a long time as far as robot-types are concerned, you know."

"No, they're robots. I've spent all day with them and I know. They've got positronic brains: primitive, of course." He placed the map in his pocket. "Let's go down."

The robots were on the lowest sublevel – all six of them surrounded by musty packing cases of uncertain content. They were large, extremely so, and even though they were in a sitting position on the floor, legs straddled out before them, their heads were a good seven feet in the air.

Donovan whistled. "Look at the size of them, will you? The chests must be ten feet around."

"That's because they're supplied with the old McGuffy gears. I've been over the insides – crummiest set you've ever seen."

"Have you powered them yet?"

"No. There wasn't any reason to. I don't think there's anything wrong with them. Even the diaphragm is in reasonable order. They might talk."

He had unscrewed the chest plate of the nearest as he spoke, inserted the two-inch sphere that contained the tiny spark of atomic energy that was a robot's life. There was difficulty in fitting it, but he managed, and then screwed the plate back on again in laborious fashion. The radio controls of more modern models had not been heard of ten years earlier. And then to the other five.

Donovan said uneasily, "They haven't moved."

"No orders to do so," replied Powell, succinctly. He went back to the first in the line and struck him on the chest. "You! Do you hear me?"

The monster's head bent slowly and the eyes fixed themselves on Powell. Then, in a harsh, squawking voice – like that of a medieval phonograph, he grated, "Yes, Master!"

Powell grinned humorlessly at Donovan. "Did you get that? Those were the days of the first talking robots when it looked as if the use of robots on Earth would be banned. The makers were fighting that and they built good, healthy slave complexes into the damned machines."

"It didn't help them," muttered Donovan.

"No, it didn't, but they sure tried." He turned once more to the robot. "Get up!"

The robot towered upward slowly and Donovan's head craned and his puckered lips whistled.

Powell said: "Can you go out upon the surface? In the light?"

There was consideration while the robot's slow brain worked. Then, "Yes, Master."

"Good. Do you know what a mile is?"

Another consideration, and another slow answer. "Yes, Master."

"We will take you up to the surface then, and indicate a direction. You will go about seventeen miles, and somewhere in that general region you will meet another robot, smaller than yourself. You understand so far?"

"Yes, Master."

"You will find this robot and order him to return. If he does not wish to, you are to bring him back by force."

Donovan clutched at Powell's sleeve. "Why not send him for the selenium direct?"

"Because I want Speedy back, nitwit. I want to find out what's wrong with him." And to the robot, "All right, you, follow me."

The robot remained motionless and his voice rumbled: "Pardon, Master, but I cannot. You must mount first." His clumsy arms had come together with a thwack, blunt fingers interlacing.

Powell stared and then pinched at his mustache. "Uh ... oh!"

Donovan's eyes bulged. "We've got to ride him? Like a horse?"

"I guess that's the idea. I don't know why, though. I can't see.–Yes, I do. I told you they were playing up robot-safety in those days. Evidently, they were going to sell the notion of safety by not allowing them to move about, without a mahout on their shoulders all the time. What do we do now?"

"That's what I've been thinking," muttered Donovan. "We can't go out on the surface, with a robot or without. Oh, for the love of Pete," – and he snapped his fingers twice. He grew excited. "Give me that map you've got. I haven't studied it for two hours for nothing. This is a Mining Station. What's wrong with using the tunnels?"

The Mining Station was a black circle on the map, and the light dotted lines that were tunnels stretched out about it in spider web fashion.

Donovan studied the list of symbols at the bottom of the map. "Look," he said, "the small black dots are openings to the surface, and here's one maybe three miles away from the selenium pool. There's a number here – you'd think they'd write larger – 13a. If the robots know their way around here – "

Powell shot the question and received the dull "Yes, Master," in reply. "Get your insosuit," he said with satisfaction.

It was the first time either had worn the insosuits – which marked one time more than either had expected to upon their arrival the day before – and they tested their limb movements uncomfortably.

The insosuit was far bulkier and far uglier than the regulation spacesuit; but withal considerably lighter, due to the fact that they were entirely nonmetallic in composition. Composed of heat-resistant plastic and chemically treated cork layers, and equipped with a desiccating unit to keep the air bone-dry, the insosuits could withstand the full glare of Mercury's sun for twenty minutes. Five to ten minutes more, as well, without actually killing the occupant.

And still the robot's hands formed the stirrup, nor did he betray the slightest atom of surprise at the grotesque figure into which Powell had been converted.

Powell's radio-harshened voice boomed out: "Are you ready to take us to Exit 13a?"

"Yes, Master."

Good, thought Powell; they might lack radio control but at least they were fitted for radio reception. "Mount one or the other, Mike," he said to Donovan.

He placed a foot in the improvised stirrup and swung upward. He found the seat comfortable; there was the humped back of the robot, evidently shaped for the purpose, a shallow groove along each shoulder for the thighs and two elongated "ears" whose purpose now seemed obvious.

Powell seized the ears and twisted the head. His mount turned ponderously. "Lead on, Macduff." But he did not feel at all lighthearted.

The gigantic robots moved slowly, with mechanical precision, through the doorway that cleared their heads by a scant foot, so that the two men had to duck hurriedly, along a narrow corridor in which their unhurried footsteps boomed monotonously and into the air lock.

The long, airless tunnel that stretched to a pinpoint before them brought home forcefully to Powell the exact magnitude of the task accomplished by the First Expedition, with their crude robots and their start-from-scratch necessities. They might have been a failure, but their failure was a good deal better than the usual run of the System's successes.

The robots plodded onward with a pace that never varied and with footsteps that never lengthened.

Powell said: "Notice that these tunnels are blazing with lights and that the temperature is Earth-normal. It's probably been like this all the ten years that this place has remained empty."

"How's that?"

"Cheap energy; cheapest in the System. Sunpower, you know, and on Mercury's Sunside, sunpower is something. That's why the Station was built in the sunlight rather than in the shadow of a mountain. It's really a huge energy converter. The heat is turned into electricity, light, mechanical work and what have you; so that energy is supplied and the Station is cooled in a simultaneous process."

"Look," said Donovan. "This is all very educational, but would you mind changing the subject? It so happens that this conversion of energy that you talk about is carried on by the photocell banks mainly – and that is a tender subject with me at the moment."

Powell grunted vaguely, and when Donovan broke the resulting silence, it was to change the subject completely. "Listen, Greg. What the devil's wrong with Speedy, anyway? I can't understand it."

It's not easy to shrug shoulders in an insosuit, but Powell tried it. "I don't know, Mike. You know he's perfectly adapted to a Mercurian environment. Heat doesn't mean anything to him and he's built for the light gravity and the broken ground. He's foolproof – or, at least, he should be."

Silence fell. This time, silence that lasted.

"Master," said the robot, "we are here."

"Eh?" Powell snapped out of a semidrowse. "Well, get us out of here – out to the surface."

They found themselves in a tiny substation, empty, airless, ruined. Donovan had inspected a jagged hole in the upper reaches of one of the walls by the light of his pocket flash.

"Meteorite, do you suppose?" he had asked.

Powell shrugged. "To hell with that. It doesn't matter. Let's get out."

A towering cliff of a black, basaltic rock cut off the sunlight, and the deep night shadow of an airless world surrounded them. Before them, the shadow reached out and ended in knife-edge abruptness into an all-but-unbearable blaze of white light that glittered from myriad crystals along a rocky ground.

"Space!" gasped Donovan. "It looks like snow." And it did.

Powell's eyes swept the jagged glitter of Mercury to the horizon and winced at the gorgeous brilliance.

"This must be an unusual area," he said. "The general albedo of Mercury is low and most of the soil is gray pumice. Something like the Moon, you know. Beautiful, isn't it?"

He was thankful for the light filters in their visiplates. Beautiful or not, a look at the sunlight through straight glass would have blinded them inside of half a minute.

Donovan was looking at the spring thermometer on his wrist. "Holy smokes, the temperature is eighty centigrade!"

Powell checked his own and said: "Um-m-m. A little high. Atmosphere, you know."

"On Mercury? Are you nuts?"

"Mercury isn't really airless," explained Powell, in absentminded fashion. He was adjusting the binocular attachments to his visiplate, and the bloated fingers of the insosuit were clumsy at it. "There is a thin exhalation that clings to its surface – vapors of the more volatile elements and compounds that are heavy enough for Mercurian gravity to retain. You know: selenium, iodine, mercury, gallium, potassium, bismuth, volatile oxides. The vapors sweep into the shadows and condense, giving up heat. It's a sort of gigantic still. In fact, if you use your flash, you'll probably find that the side of the cliff is covered with, say, hoar-sulphur, or maybe quicksilver dew.

"It doesn't matter, though. Our suits can stand a measly eighty indefinitely."

Powell had adjusted the binocular attachments, so that he seemed as eye-stalked as a snail.

Donovan watched tensely. "See anything?"

The other did not answer immediately, and when he did, his voice was anxious and thoughtful. "There's a dark spot on the horizon that might be the selenium pool. It's in the right place. But I don't see Speedy."

Powell clambered upward in an instinctive striving for better view, till he was standing in unsteady fashion upon his robot's shoulders. Legs straddled wide, eyes straining, he said: "I think ... I think – Yes, it's definitely he. He's coming this way."

Donovan followed the pointing finger. He had no binoculars, but there was a tiny moving dot, black against the blazing brilliance of the crystalline ground.

"I see him," he yelled. "Let's get going!"

Powell had hopped down into a sitting position on the robot again, and his suited hand slapped against the Gargantuan's barrel chest. "Get going!"

"Giddy-ap," yelled Donovan, and thumped his heels, spur fashion.

The robots started off, the regular thudding of their footsteps silent in the airlessness, for the nonmetallic fabric of the insosuits did not transmit sound. There was only a rhythmic vibration just below the border of actual hearing.

"Faster," yelled Donovan. The rhythm did not change.

"No use," cried Powell, in reply. "These junk heaps are only geared to one speed. Do you think they're equipped with selective flexors?"

They had burst through the shadow, and the sunlight came down in a white-hot wash and poured liquidly about them.

Donovan ducked involuntarily. "Wow! Is it imagination or do I feel heat?"

"You'll feel more presently," was the grim reply. "Keep your eye on Speedy."

Robot SPD 13 was near enough to be seen in detail now. His graceful, streamlined body threw out blazing highlights as he loped with easy speed across the broken ground. His name was derived from his serial initials, of course, but it was apt, nevertheless, for the SPD models were among the fastest robots turned out by the United States Robot & Mechanical Men Corp.

"Hey, Speedy," howled Donovan, and waved a frantic hand.

"Speedy!" shouted Powell. "Come here!"

The distance between the men and the errant robot was being cut down momentarily – more by the efforts of Speedy than the slow plodding of the fifty-year-old antique mounts of Donovan and Powell.

They were close enough now to notice that Speedy's gait included a peculiar rolling stagger, a noticeable side-to-side lurch – and then, as Powell waved his hand again and sent maximum juice into his compact headset radio sender, in preparation for another shout, Speedy looked up and saw them.

Speedy hopped to a halt and remained standing for a moment with just a tiny, unsteady weave, as though he were swaying in a light wind.

Powell yelled: "All right, Speedy. Come here, boy."

Whereupon Speedy's robot voice sounded in Powell's earphones for the first time.

It said: "Hot dog, let's play games. You catch me and I catch you; no love can cut our knife in two. For I'm Little Buttercup, sweet Little Buttercup. Whoops!" Turning on his heel, he sped off in the direction from which he had come, with a speed and fury that kicked up gouts of baked dust.

And his last words as he receded into the distance were, "There grew a little flower 'neath a great oak tree," followed by a curious metallic clicking that might have been a robotic equivalent of a hiccup.

Donovan said weakly: "Where did he pick up the Gilbert and Sullivan? Say, Greg, he ... he's drunk or something."

"If you hadn't told me," was the bitter response, "I'd never realize it. Let's get back to the cliff. I'm roasting."

It was Powell who broke the desperate silence. "In the first place," he said, "Speedy isn't drunk – not in the human sense – because he's a robot, and robots don't get drunk. However, there's something wrong with him which is the robotic equivalent of drunkenness."

"To me, he's drunk," stated Donovan, emphatically, "and all I know is that he thinks we're playing games. And we're not. It's a matter of life and very gruesome death."

"All right. Don't hurry me. A robot's only a robot. Once we find out what's wrong with him, we can fix it and go on."

"Once," said Donovan, sourly.

Powell ignored him. "Speedy is perfectly adapted to normal Mercurian environment. But this region" – and his arm swept wide – "is definitely abnormal. There's our clue. Now where do these crystals come from? They might have formed from a slowly cooling liquid; but where would you get liquid so hot that it would cool in Mercury's sun?"

"Volcanic action," suggested Donovan, instantly, and Powell's body tensed.

"Out of the mouths of sucklings," he said in a small, strange voice and remained very still for five minutes.

Then, he said, "Listen, Mike, what did you say to Speedy when you sent him after the selenium?"

Donovan was taken aback. "Well damn it – I don't know. I just told him to get it."

"Yes, I know, but how? Try to remember the exact words."

"I said ... uh ... I said: 'Speedy, we need some selenium. You can get it such-and-such a place. Go get it – that's all. What more did you want me to say?"

"You didn't put any urgency into the order, did you?"

"What for? It was pure routine."

Powell sighed. "Well, it can't be helped now – but we're in a fine fix." He had dismounted from his robot, and was sitting back against the cliff. Donovan joined him and they linked arms: In the distance the burning sunlight seemed to wait cat-and-mouse for them, and just next them, the two giant robots were invisible but for the dull red of their photoelectric eyes that stared down at them, unblinking, unwavering, and unconcerned.

Unconcerned! As was all this poisonous Mercury, as large in jinx as it was small in size.

Powell's radio voice was tense in Donovan's ear: "Now, look, let's start with the three fundamental Rules of Robotics – the three rules that are built most deeply into a robot's positronic brain." In the darkness, his gloved fingers ticked off each point.

"We have: One, a robot may not injure a human being, or, through inaction, allow a human being to come to harm."

"Right!"

"Two," continued Powell, "a robot must obey the orders given it by human beings except where such orders would conflict with the First Law."

"Right."

"And three, a robot must protect its own existence as long as such protection does not conflict with the First or Second Laws."

"Right! Now where are we?"

"Exactly at the explanation. The conflict between the various rules is ironed out by the different positronic potentials in the brain. We'll say that a robot is walking into danger and knows it. The automatic potential that Rule 3 sets up turns him back. But suppose you order him to walk into that danger. In that case, Rule 2 sets up a counterpotential higher than the previous one and the robot follows orders at the risk of existence."

"Well, I know that. What about it?"

"Let's take Speedy's case. Speedy is one of the latest models, extremely specialized, and as expensive as a battleship. It's not a thing to be lightly destroyed."

"So?"

"So Rule 3 has been strengthened – that was specifically mentioned, by the way, in the advance notices on the SPD models – so that his allergy to danger is unusually high. At the same time, when you sent him out after the selenium, you gave him his order casually and without special emphasis, so that the Rule 2 potential set-up was rather weak. Now, hold on; I'm just stating facts."

"All right, go ahead. I think I get it."

"You see how it works, don't you? There's some sort of danger centering at the selenium pool. It increases as he approaches, and at a certain distance from it the Rule 3 potential, unusually high to start with, exactly balances the Rule 2 potential, unusually low to start with."

Donovan rose to his feet in excitement. "And it strikes an equilibrium. I see. Rule 3 drives him back and Rule 2 drives him forward – "

"So he follows a circle around the selenium pool, staying on the locus of all points of potential equilibrium. And unless we do something about it, he'll stay on that circle forever, giving us the good old runaround." Then, more thoughtfully: "And that, by the way, is what makes him drunk. At potential equilibrium, half the positronic paths of his brain are out of kilter. I'm not a robot specialist, but that seems obvious. Probably he's lost control of just those parts of his voluntary mechanism that a human drunk has. Ve-e-ery pretty."

"But what's the danger? If we knew what he was running from – ?"

"You suggested it. Volcanic action. Somewhere right above the selenium pool is a seepage of gas from the bowels of Mercury. Sulphur dioxide, carbon dioxide – and carbon monoxide. Lots of it and at this temperature."

Donovan gulped audibly. "Carbon monoxide plus iron gives the volatile iron carbonyl."

"And a robot," added Powell, "is essentially iron." Then, grimly: "There's nothing like deduction. We've determined everything about our problem but the solution. We can't get the selenium ourselves. It's still too far. We can't send these robot horses, because they can't go themselves, and they can't carry us fast enough to keep us from crisping. And we can't catch Speedy, because the dope thinks we're playing games, and he can run sixty miles to our four."

"If one of us goes," began Donovan, tentatively, "and comes back cooked, there'll still be the other."

"Yes," came the sarcastic reply, "it would be a most tender sacrifice – except that a person would be in no condition to give orders before he ever reached the pool, and I don't think the robots would ever turn back to the cliff without orders. Figure it out! We're two or three miles from the pool – call it two – the robot travels at four miles an hour; and we can last twenty minutes in our suits. It isn't only the heat, remember. Solar radiation out here in the ultraviolet and below is poison."

"Um-m-m," said Donovan, "ten minutes short."

"As good as an eternity. And another thing, in order for Rule 3 potential to have stopped Speedy where it did, there must be an appreciable amount of carbon monoxide in the metal-vapor atmosphere – and there must be an appreciable corrosive action therefore. He's been out hours now – and how do we know when a knee joint, for instance, won't be thrown out of kilter and keel him over. It's not only a question of thinking – we've got to think fast!"

Deep, dark, dank, dismal silence!

Donovan broke it, voice trembling in an effort to keep itself emotionless. He said: "As long as we can't increase Rule 2 potential by giving further orders, how about working the other way? If we increase the danger, we increase Rule 3 potential and drive him backward."

Powell's visiplate had turned toward him in a silent question.

"You see," came the cautious explanation, "all we need to do to drive him out of his rut is to increase the concentration of carbon monoxide in his vicinity. Well, back at the Station there's a complete analytical laboratory."

"Naturally," assented Powell. "It's a Mining Station."

"All right. There must be pounds of oxalic acid for calcium precipitations."

"Holy space! Mike, you're a genius."

"So-so," admitted Donovan, modestly. "It's just a case of remembering that oxalic acid on heating decomposes into carbon dioxide, water, and good old carbon monoxide. College chem, you know."

Powell was on his feet and had attracted the attention of one of the monster robots by the simple expedient of pounding the machine's thigh.

"Hey," he shouted, "can you throw?"

"Master?"

"Never mind." Powell damned the robot's molasses-slow brain. He scrabbled up a jagged brick-size rock. "Take this," he said, "and hit the patch of bluish crystals just across the crooked fissure. You see it?"

Donovan pulled at his shoulder. "Too far, Greg. It's almost half a mile off."

"Quiet," replied Powell. "It's a case of Mercurian gravity and a steel throwing arm. Watch, will you?"

The robot's eyes were measuring the distance with machinely accurate stereoscopy. His arm adjusted itself to the weight of the missile and drew back. In the darkness, the robot's motions went unseen, but there was a sudden thumping sound as he shifted his weight, and seconds later the rock flew blackly into the sunlight. There was no air resistance to slow it down, nor wind to turn it aside – and when it hit the ground it threw up crystals precisely in the center of the "blue patch."

Powell yelled happily and shouted, "Let's go back after the oxalic acid, Mike."

And as they plunged into the ruined substation on the way back to the tunnels, Donovan said grimly: "Speedy's been hanging about on this side of the selenium pool, ever since we chased after him. Did you see him?"

"Yes."

"I guess he wants to play games. Well, we'll play him games!"

They were back hours later, with three-liter jars of the white chemical and a pair of long faces. The photocell banks were deteriorating more rapidly than had seemed likely. The two steered their robots into the sunlight and toward the waiting Speedy in silence and with grim purpose.

Speedy galloped slowly toward them. “Here we are again. Whee! I’ve made a little list, the piano organist; all people who eat peppermint and puff it in your face.”

“We’ll puff something in your face,” muttered Donovan. “He’s limping, Greg.”

“I noticed that,” came the low, worried response. “The monoxide’ll get him yet, if we don’t hurry.”

They were approaching cautiously now, almost sidling, to refrain from setting off the thoroughly irrational robot. Powell was too far off to tell, of course, but even already he could have sworn the crack-brained Speedy was setting himself for a spring.

“Let her go,” he gasped. “Count three! One – two – ”

Two steel arms drew back and snapped forward simultaneously and two glass jars whirled forward in towering parallel arcs, gleaming like diamonds in the impossible sun. And in a pair of soundless puffs, they hit the ground behind Speedy in crashes that sent the oxalic acid flying like dust.

In the full heat of Mercury’s sun, Powell knew it was fizzing like soda water.

Speedy turned to stare, then backed away from it slowly – and as slowly gathered speed. In fifteen seconds, he was leaping directly toward the two humans in an unsteady canter.

Powell did not get Speedy’s words just then, though he heard something that resembled, “Lover’s professions when uttered in Hessians.”

He turned away. “Back to the cliff, Mike. He’s out of the rut and he’ll be taking orders now. I’m getting hot.”

They jogged toward the shadow at the slow monotonous pace of their mounts, and it was not until they had entered it and felt the sudden coolness settle softly about them that Donovan looked back. “Greg!”

Powell looked and almost shrieked. Speedy was moving slowly now – so slowly – and in the wrong direction. He was drifting; drifting back into his rut; and he was picking up speed. He looked dreadfully close, and dreadfully unreachable, in the binoculars.

Donovan shouted wildly, “After him!” and thumped his robot into its pace, but Powell called him back.

“You won’t catch him, Mike – it’s no use.” He fidgeted on his robot’s shoulders and clenched his fist in tight impotence. “Why the devil do I see these things five seconds after it’s all over? Mike, we’ve wasted hours.”

“We need more oxalic acid,” declared Donovan, stolidly. “The concentration wasn’t high enough.”

“Seven tons of it wouldn’t have been enough – and we haven’t the hours to spare to get it, even if it were, with the monoxide chewing him away. Don’t you see what it is, Mike?”

And Donovan said flatly, “No.”

“We were only establishing new equilibriums. When we create new monoxide and increase Rule 3 potential, he moves backward till he’s in balance again – and when the monoxide drifted away, he moved forward, and again there was balance.”

Powell’s voice sounded thoroughly wretched. “It’s the same old runaround. We can push at Rule 2 and pull at Rule 3 and we can’t get anywhere – we can only change the position of balance. We’ve got to get outside both rules.” And then he pushed his robot closer to Donovan’s so that they were sitting face-to-face, dim shadows in the darkness, and he whispered, “Mike!”

“Is it the finish?” – dully. “I suppose we go back to the Station, wait for the banks to fold, shake hands, take cyanide, and go out like gentlemen.” He laughed shortly.

“Mike,” repeated Powell earnestly, “we’ve got to get Speedy.”

“I know.”

“Mike,” once more, and Powell hesitated before continuing. “There’s always Rule 1. I thought of it – earlier – but it’s desperate.”

Donovan looked up and his voice livened. “We’re desperate.”

“All right. According to Rule 1, a robot can’t see a human come to harm because of his own inaction. Two and 3 can’t stand against it. They can’t, Mike.”

“Even when the robot is half cra – Well, he’s drunk. You know he is.”

“It’s the chances you take.”

“Cut it. What are you going to do?”

“I’m going out there now and see what Rule 1 will do. If it won’t break the balance, then what the devil – it’s either now or three-four days from now.”

“Hold on, Greg. There are human rules of behavior, too. You don’t go out there just like that. Figure out a lottery, and give me my chance.”

“All right. First to get the cube of fourteen goes.” And almost immediately, “Twenty-seven forty-four!”

Donovan felt his robot stagger at a sudden push by Powell’s mount and then Powell was off into the sunlight. Donovan opened his mouth to shout, and then clicked it shut. Of course, the damn fool had worked out the cube of fourteen in advance, and on purpose. Just like him.

The sun was hotter than ever and Powell felt a maddening itch in the small of his back. Imagination, probably, or perhaps hard radiation beginning to tell even through the insosuit.

Speedy was watching him, without a word of Gilbert and Sullivan gibberish as greeting. Thank God for that! But he daren’t get too close.

He was three hundred yards away when Speedy began backing, a step at a time, cautiously – and Powell stopped. He jumped from his robot’s shoulders and landed on the crystalline ground with a light thump and a flying of jagged fragments.

He proceeded on foot, the ground gritty and slippery to his steps, the low gravity causing him difficulty. The soles of his feet tickled with warmth. He cast one glance over his shoulder at the blackness of the cliff's shadow and realized that he had come too far to return – either by himself or by the help of his antique robot. It was Speedy or nothing now, and the knowledge of that constricted his chest.

Far enough! He stopped.

"Speedy," he called. "Speedy!"

The sleek, modern robot ahead of him hesitated and halted his backward steps, then resumed them.

Powell tried to put a note of pleading into his voice, and found it didn't take much acting. "Speedy, I've got to get back to the shadow or the sun'll get me. It's life or death, Speedy. I need you."

Speedy took one step forward and stopped. He spoke, but at the sound Powell groaned, for it was, "When you're lying awake with a dismal headache and repose is tabooed – " It trailed off there, and Powell took time out for some reason to murmur, "Iolanthe."

It was roasting hot! He caught a movement out of the corner of his eye, and whirled dizzily; then stared in utter astonishment, for the monstrous robot on which he had ridden was moving – moving toward him, and without a rider.

He was talking: "Pardon, Master. I must not move without a Master upon me, but you are in danger."

Of course, Rule 1 potential above everything. But he didn't want that clumsy antique; he wanted Speedy. He walked away and motioned frantically: "I order you to stay away. I order you to stop!"

It was quite useless. You could not beat Rule 1 potential. The robot said stupidly, "You are in danger, Master."

Powell looked about him desperately. He couldn't see clearly. His brain was in a heated whirl; his breath scorched when he breathed, and the ground all about him was a shimmering haze.

He called a last time, desperately: "Speedy! I'm dying, damn you! Where are you? Speedy, I need you."

He was still stumbling backward in a blind effort to get away from the giant robot he didn't want, when he felt steel fingers on his arms, and a worried, apologetic voice of metallic timbre in his ears.

"Holy smokes, boss; what are you doing here? And what am I doing – I'm so confused – "

"Never mind," murmured Powell, weakly. "Get me to the shadow of the cliff – and hurry!" There was one last feeling of being lifted into the air and a sensation of rapid motion and burning heat, and he passed out.

He woke with Donovan bending over him and smiling anxiously. "How are you, Greg?"

"Fine!" came the response, "Where's Speedy?"

"Right here. I sent him out to one of the other selenium pools – with orders to get that selenium at all cost this time. He got it back in forty-two minutes and three seconds. I timed him. He still hasn't finished apologizing for the runaround he gave us. He's scared to come near you for fear of what you'll say."

"Drag him over," ordered Powell. "It wasn't his fault." He held out a hand and gripped Speedy's metal paw. "It's O.K., Speedy." Then, to Donovan, "You know, Mike, I was just thinking – "

"Yes!"

"Well," – he rubbed his face – the air was so delightfully cool, "you know that when we get things set up here and Speedy put through his Field Tests, they're going to send us to the Space Stations next – "

"No!"

"Yes! At least that's what old lady Calvin told me just before we left, and I didn't say anything about it, because I was going to fight the whole idea."

"Fight it?" cried Donovan. "But – "

"I know. It's all right with me now. Two hundred seventy-three degrees Centigrade below zero. Won't it be a pleasure?"

"Space Station," said Donovan, "here I come."

After You Have Read the Story …

"Runaround" is an amusing exploration of the inherent, hidden conflicts in the Three Laws of Robotics. It also serves to illustrate how a robot can be comprised of independent behavioral impulses, where each imposes a "force" or vector on the robot. The independent vectors produce a single resultant vector that describes the direction and velocity of the robot's actual movement. In the story, vectors from multiple behaviors can produce unintended consequences but in practice, it is easy to simulate the emergent vector and detect surprises. There are also several mechanisms that roboticists use to eliminate surprises. For more in-depth reading, chapter 6 in *Introduction to AI Robotics* introduces behaviors and chapter 8 describes potential field methodologies in detail, and for advanced reading, Ron Arkin's *Behavior-Based Robotics* is an excellent source.[2]

Behaviors

The introduction of behaviors in robots and the use of potential field methodologies as one way of representing the output of a behavior started in the mid-1980s. At that time, artificial intelligence researchers turned to biology for insights into creating intelligence. Advances in robotics had been slow in coming after the initial success in 1967 of the Shakey project (introduced in chapter 1).[3] Shakey evinced a generally accepted paradigm that an intelligent robot should first sense the world, construct an internal symbol model of the world from the sensor readings and what it already knew, plan the optimal next action to accomplish its goal, and then execute that action. This was called SENSE-PLAN-ACT and formed a monolithic program in which a robot sensed, then planned, and then acted in a continuous loop. The sequence appeared to be logical and appealed to the military, and especially to pilots who were trained to think in a very similar Observe, Orient, Detect, and Act (OODA) loop pattern.

Unfortunately, sensing and planning were both very hard. Computer vision was unable to recognize even simple objects except in controlled lighting conditions. No one knew what the right categories or labels for different objects in the world should be. Planning an optimal action such as a path through a room with a few chairs in the way could take hours of computation.

Stymied by the lack of progress and motivated by engineering researchers such as Michael Arbib who had begun to apply engineering analysis methods to animals, a new cadre of researchers including Rod Brooks, Ron Arkin, and David Payton began to influence robotics, creating a new field of behavioral robotics summarized in *Behavior-Based Robots*.[4] They examined how animals were able to successfully move through complex environments with little or no apparent computation. Brooks went even further and built robots that looked like the insects that used such simple behaviors.

What that first generation of behavioral roboticists found was that animal intelligence is built on strata of simple behaviors. Animals don't SENSE-PLAN-ACT; they sense a predator and immediately act by running away, with no planning in between. A toad or a frog will sense that something is in its way, but not categorize or recognize it, and simply move to avoid it. Animals react to the world; they don't plan. These SENSE-ACT units are called behaviors and can be either hardwired or learned.

It isn't just "dumb" animals such as fish and frogs that rely on SENSE-ACT behaviors—people do the same thing. During World War II, the cognitive psychologist J. J. Gibson found that pilots reacted faster and more reliably to depth cues in the environment, such as how fast a plane appeared to be moving and changes in altitude, than looking at gauges and dials indicating altitude and aircraft.[5] As a result, painting lines and having lights on runways at regular intervals was the biggest help to pilots landing aircraft; they didn't have to think about landing, they could just revert to hand-eye (SENSE-ACT) coordination skills. The conclusion in psychology was that we humans, like other animals, have a whole host of behaviors—whether hardwired or learned—that, when deployed, associate what we are sensing with how we should react.

Arbib's studies with neuroscientists found that the behavior of toads could be represented as a set of behaviors using potential fields.[6] A toad being attracted to a fly can be set up as a field, like a gravitational field, that would attract or pull the toad to the fly. At the same time, any obstacles in the way would create a separate repulsive field that would push the toad away. A toad would "feel" a vector that pulled it toward the fly and at the same time "feel" another vector that pushed it away from the obstacle—these vectors would be summed in the brain and produce a resultant vector.

The toad would hop slightly toward the fly but to the side of the obstacle, and then repeat the process.

The insight that behaviors could be represented by simple repulsive or attractive vectors led to breakthroughs in robotics. A robot such as a Roomba vacuum cleaner no longer had to understand the difference between a wall and a chair, just be repulsed by whatever was in its way. Although planning is an important part of some intelligent functions, such as scheduling and allocating resources or performing complex assemblages of parts, it isn't necessary for every intelligent function.

This perception was viewed as heresy by other researchers who wanted robotics to be perfectly analytical and robots to be "better" or "purer" than animals. After two decades of infighting among researchers, the field has generally converged on an understanding that some activities can be implemented strictly with reactive behaviors (called behavior-based robotics) but other activities require more sophisticated, deliberative functions that involve planning and understanding the world in addition to those reactive behaviors.

Five Primitive Potential Fields for Representing Behavioral Output

By the 1990s, researchers had created five primitive potential fields from which all reactive behaviors could be constructed: *uniform, perpendicular, attraction, repulsion,* and *tangential.* These are shown in figure 2.2 and an example of how they would be combined is shown in figure 2.3. At this time, researchers had also begun to explore interesting effects like *gains.* If an animal was very, very hungry, the attraction vector to prey might be literally doubled by a multiplier, or gain, on the vector magnitude. Behavior-based robotics remains a rich subdiscipline of inquiry to this day.

The Local Minima Problem

But there was a fly in the ointment—mathematically, vectors can sum to zero, a situation that is called a *local minima.* In physics, this would occur when a metal ball is equidistant from two magnets and thus perfectly balanced. But in intelligence, it means that situations could arise in which the vectors from competing behaviors result in the robot doing nothing. These situations rarely occur in nature as animals have noisy sensors and a new, slightly different sensor update will change their response: another behavioral input such as hunger will kick in, or something in the environment

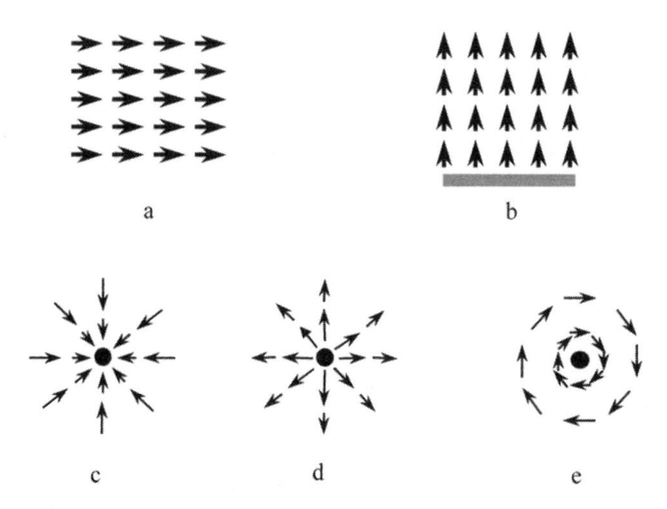

Figure 2.2
Five primitive potential fields: a) uniform, b) perpendicular, c) attraction, d) repulsion, and e) tangential (from Murphy, *Introduction to AI Robotics*).

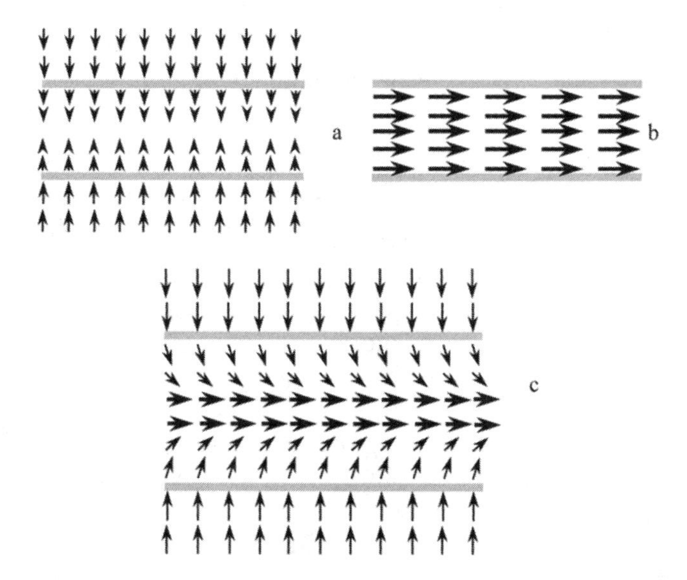

Figure 2.3
(a) Perpendicular fields and (b) uniform fields combine in a (c) follow-corridor behavior (from Murphy, *Introduction to AI Robotics*).

will change, such as a predator's location. The possibility that a local minima could occur in robots, however, was quite a cause of concern. Significant effort was devoted to mechanisms for ensuring that a local minima would never occur, including adding random noise to the system or representing the potential fields with harmonic functions. As a result, local minima is not a real problem for a well-engineered system.

Whereas local minima is generally not a problem for actual behavior-based robotic systems, it is for Speedy. He is experiencing both an attractive vector to the selenium pool because he has been ordered to get selenium and also some sort of repulsive vector for self-protection.

One question this raises is why is Speedy circling the selenium pool given the vectors? If the attractive and repulsive fields originate from the center of the pool, the vectors should sum to zero and Speedy should be totally still. Instead, the story reads as if the selenium pool produces a tangential field that wants to "swirl" Speedy away from danger rather than a strictly repulsive or "go backward" field. If there were a tangential field, then that would explain the drunken walk: Speedy would turn and go at a tangent to the pool, and then the distance between Speedy and the pool would increase to a safe level, so Speedy would turn back toward the pool, and the cycle would repeat.

The story also touches on the idea of gains. It implies that in a normal situation, the behaviors would be less evenly weighted and thus Speedy would have been able to get close enough to acquire the selenium.

Reality Score: B+

Although implementing behaviors as potential fields is realistic, "Runaround" mixes reactive functions with deliberative functions (deliberation will be discussed in the next chapter). Speedy is too dumb to notice that it is trapped but smart enough to know Gilbert and Sullivan? Not likely!

3 Deliberation: "Long Shot"

It should be no surprise that Vernor Vinge writes highly plausible science fiction, as he is a computer scientist and former professor of mathematics at San Diego State University.[1] Vinge is famous for posing the idea of the singularity, which influenced Hans Moravec's projection that following advances described by Moore's law, people, computers, and robots would soon merge and become transhumans, as described in *Mind Children*.[2]

The plot of "Long Shot," written in 1972, is similar to the classic Heinlein story "Universe" in which a spaceship has been in space so long and working under such unfavorable conditions that it does not remember its mission.[3] In "Universe," it is up to the human (and post-human) crewmembers to rediscover the goal of the mission and resume progress after significant damage. In "Long Shot," the technological MacGuffin is that it is up to the robot ship Ilse herself to reach the goal of her mission. Unlike "Universe," she doesn't start with enough memory to have the ultimate goal loaded in, mirroring the early days of computers when files were stored on sets of magnetic tape. But along the way, Ilse has to handle a "Universe"-like major challenge that real spaceships have to handle: unpredictable degradation and damage to her hardware and memory.

Deep Space One, a NASA probe, highlights how AI can monitor for damage and generate repairs or workarounds. It was launched in 1998 to fly by an asteroid and comet Borrelly. In order to reliably complete its mission through 2001, it used a technique called model-based diagnosis.[4] Although "Long Shot" doesn't list specific algorithms, it gets everything right about the use of AI and robotics to remember and execute a mission.

As You Read the Story ...

Ilse is a fully autonomous agent, traveling in space beyond any possible communication link with a human supervisor. The technological challenge is how an agent can be resilient over time—sometimes called *long-term autonomy*—as the designers cannot anticipate every possible failure mode or situation it might encounter. The inability to model every possibility is called the *open world assumption* in artificial intelligence. Industrial or factory robots operate in a "closed world" (or the *closed world assumption*) where the environment is engineered for the robot and there should be no surprises. In a closed world, the robot can't sense or comprehend anything or any situation that isn't explicitly entered in its knowledge base. If a worker enters the robot's workspace while it is operating, the robot will keep performing the planned actions and thus may accidently harm or kill the person.

The story serves as an introduction to deliberation, which is covered more fully in chapter 12 of *Introduction to AI Robotics*.[5] As you read, keep track of the different ways in which Ilse plans and solves problems. These will generally fall into four categories: *generating* a plan, *selecting* and scheduling resources, *implementing* the plan with the resources, and *monitoring* the plan's execution. Also think about whether she is using a closed or open world assumption for each situation. The "After You Have Read the Story" section will review the open and closed world assumptions, introduce four categories of deliberation, discuss how deliberation relates to time horizons, and present the general software architecture used in AI robotics to capture behavioral and deliberative functions.

"Long Shot" by Vernor Vinge, 1972

By itself, it seems unlikely that war could destroy the human race, or even bring a permanent halt to our slide into the Singularity. Yet the universe itself can be a rough place; we have plenty of evidence of mass extinctions. If we had a technology-smashing war *plus* an extended natural catastrophe, we could join the dinosaurs.

And of course, there are natural cataclysms that can destroy not just life but entire planets. Fortunately, the most extreme stellar catastrophes – such as supernovas – are impossible for a star like our sun. What about smaller events, burps in the life of otherwise placid stars? We have no guarantee that our sun is safe from these. What would we do if, in the next fifteen years, we discovered that our sun was about to enter an extended period of increased luminance, frying the surfaces of the inner planets? Given a decade, could we establish a self-sustaining colony in the outer solar system? If not, could we find Earth-like planets elsewhere? At present, sending even the smallest probe to the nearest stars is just beyond our ability. Not a single living person could be saved. Whatever we tried would indeed be a ...

LONG SHOT

They named her Ilse, and of all Earth's creatures, she was to be the longest lived – and perhaps the last. A prudent tortoise might survive three hundred years and a bristle-cone pine six thousand, but Ilse's designed span exceeded one hundred centuries. And though her brain was iron and germanium doped with arsenic, and her heart was a tiny cloud of hydrogen plasma, Ilse *was* – in the beginning – one of Earth's creatures: she could feel, she could question, and – as she discovered during the dark centuries before her fiery end – she could also forget.

Ilse's earliest memory was a fragment, amounting to less than fifteen seconds. Someone, perhaps inadvertently, brought her to consciousness as she sat atop her S-5N booster. It was night, but their launch was imminent and the booster stood white and silver in the light of a dozen spotlights. Ilse's sharp eye scanned rapidly around the horizon, untroubled by the glare from below. Stretching away from her to the north was a line of thirty launch pads. Several had their own boosters, though none were lit up as Ilse's was. Three thousand meters to the west were more lights, and the occasional sparkle of an automatic rifle. To the east, surf marched in phosphorescent ranks against the Merritt Island shore.

There the fragment ended: she was not conscious during the launch. But that scene remained forever her most vivid and incomprehensible memory.

When next she woke, Ilse was in low Earth orbit. Her single eye had been fitted to a one hundred and fifty centimeter reflecting telescope so that now she could distinguish stars set less than a tenth of a second apart, or, if she looked straight down, count the birds in a flock of geese two hundred kilometers below. For more than a year Ilse remained in this same orbit. She was not idle. Her makers had allotted this period for testing. A small manned station orbited with her, and from it came an endless sequence of radioed instructions and exercises.

Most of the problems were ballistic: hyperbolic encounters, transfer ellipses, and the like. But it was often required that Ilse use her own telescope and spectrometer to discover the parameters of the problems. A typical exercise: determine the orbits of Venus and Mercury; compute a minimum energy flyby of both planets. Another: determine the orbit of Mars; analyze its atmosphere; plan a hyperbolic entry subject to constraints. Many observational problems dealt with Earth: determine atmospheric pressure and composition; perform multispectrum analysis of vegetation. Usually she was required to solve organic analysis problems in less than thirty seconds. And in these last problems, the rules were often changed even while the game was played. Her orientation jets would be caused to malfunction. Critical portions of her mind and senses would be degraded.

One of the first things Ilse learned was that in addition to her private memories, she had a programmed memory, a "library" of procedures and facts. As with most libraries, the programmed memory was not as accessible as Ilse's own recollections, but the information contained there was much more complete and precise. The solution program for almost any ballistic, or chemical, problem could be lifted from this "library," used for seconds, or hours, as an integral part of Ilse's mind, and then returned to the "library." The real trick was to select the proper program on the basis of incomplete information, and then to modify that program to meet various combinations of power and equipment failure. Though she did poorly at first, Ilse eventually surpassed her design specifications. At this point her training stopped and for the first – but not the last – time. Ilse was left to her own devices.

Though she had yet to wonder on her ultimate purpose, still she wanted to see as much of her world as possible. She spent most of each daylight pass looking straight down, trying to see some order in the jumble of blue and green and white. She could easily follow the supply rockets as they climbed up from Merritt Island and Baikonur to rendezvous with her. In the end, more than a hundred of the rockets were floating about her. As the weeks passed, the squat white cylinders were fitted together on a spidery frame.

Now her ten-meter-long body was lost in the webwork of cylinders and girders that stretched out two hundred meters behind her. Her programmed memory told her that the entire assembly massed 22,563.901 tons – more than most ocean-going ships – and a little experimenting with her attitude control jets convinced her that this figure was correct.

Soon her makers connected Ilse's senses to the mammoth's control mechanisms. It was as if she had been given a new body, for she could feel, and see, and use each of the hundred propellant tanks and each of the fifteen fusion reactors that made up the assembly. She realized that now she had the power to perform some of the maneuvers she had planned during her training.

Finally the great moment arrived. Course directions came over the maser link with the manned satellite. Ilse quickly computed the trajectory that would result from these directions. The answer she obtained was correct, but it revealed only the smallest part of what was in store for her.

In her orbit two hundred kilometers up, Ilse coasted smoothly toward high noon over the Pacific. Her eye was pointed forward, so that on the fuzzy blue horizon she could see the edge of the North American continent. Nearer, the granulated cloud cover obscured the ocean itself. The command to begin the burn came from the manned satellite, but Ilse was following the clock herself, and she had determined to take over the launch if any mistakes were made. Two hundred meters behind her, deep in the maze of tanks and beryllium girders, Ilse felt magnetic fields establish themselves, felt hydrogen plasma form, felt fusion commence. Another signal from the station, and propellant flowed around each of ten reactors.

Ilse and her twenty-thousand-ton booster were on their way.

Acceleration rose smoothly to one gravity. Behind her, vidicons on the booster's superstructure showed the Earth shrinking. For half an hour the burn continued, monitored by Ilse, and the manned station now fallen far behind.

Then Ilse was alone with her booster, coasting away from Earth and her creators at better than twenty kilometers a second.

So Ilse began her fall toward the sun. For eleven weeks she fell. During this time, there was little to do: monitor the propellants, keep the booster's sunshade properly oriented, relay data to Earth. Compared to much of her later life, however, it was a time of hectic activity.

A fall of eleven weeks toward a body as massive as the sun can result in only one thing: speed. In those last hours, Ilse hurtled downwards at better than two hundred and fifty kilometers per second – an Earth to Moon distance every half hour. Forty-five minutes before her closest approach to the sun – perihelion – Ilse jettisoned the empty first stage and its sunshade. Now she was left with the two-thousand-ton second stage, whose insulation consisted of a bright coat of white paint. She felt the pressure in the propellant tanks begin to rise.

Though her telescope was pointed directly away from the sun, the vidicons on the second stage gave her an awesome view of the solar fireball. She was moving so fast now that the sun's incandescent prominences changed perspective even as she watched.

Seventeen minutes to perihelion. From somewhere beyond the flames, Ilse got the expected maser communication. She pitched herself and her booster over so that she looked along the line of her trajectory. Now her own body was exposed to the direct glare of the sun. Through her telescope she could see luminous tracery within the solar corona. The booster's fuel tanks were perilously close to bursting, and Ilse was having trouble keeping her own body at its proper temperature.

Fifteen minutes to perihelion. The command came from Earth to begin the burn. Ilse considered her own trajectory data, and concluded that the command was thirteen seconds premature. Consultation with Earth would cost at least sixteen minutes, and her decision must be made in the next four seconds. Any of Man's earlier, less sophisticated creations would have accepted the error and taken the mission on to catastrophe, but independence was the essence of Ilse's nature: she overrode the maser command, and delayed ignition till the instant she thought correct.

The sun's northern hemisphere passed below her, less than three solar diameters away.

Ignition, and Ilse was accelerated at nearly two gravities. As she swung toward what was to have been perihelion, her booster lifted her out of elliptic orbit and into a hyperbolic one. Half an hour later she shot out from the sun into the spaces south of the ecliptic at three hundred and twenty kilometers per second – about one solar diameter every hour. The booster's now empty propellant tanks were between her and the sun, and her body slowly cooled.

Shortly after burnout, Earth off-handedly acknowledged the navigation error. This is not to say that Ilse's makers were without contrition for their mistake, or without praise for Ilse. In fact, several men lost what little there remained to confiscate for jeopardizing this mission, and Man's last hope. It was simply that Ilse's makers did not believe that she could appreciate apologies or praise.

Now Ilse fled up out of the solar gravity well. It had taken her eleven weeks to fall from Earth to Sol, but in less than two weeks she had regained this altitude, and still she plunged outwards at more than one hundred kilometers per second. That velocity remained her inheritance from the sun. Without the gravity-well maneuver, her booster would have had to be five hundred times as large, or her voyage three times as long. It had been the very best that men could do for her, considering the time remaining to them.

So began the voyage of one hundred centuries. Ilse parted with the empty booster and floated on alone: a squat cylinder, twelve meters wide, five meters long, with a large telescope sticking from one end. Four light-years below her in the well of the night she saw Alpha Centauri, her destination. To the naked human eye, it appears a single bright star, but with her telescope Ilse could clearly see two stars, one slightly fainter and redder than the other. She carefully measured their position and her own, and concluded that her aim had been so perfect that a midcourse correction would not be necessary for a thousand years.

For many months, Earth maintained maser contact – to pose problems and ask after her health. It was almost pathetic, for if anything went wrong now, or in the centuries to follow, there was very little Earth could do to help. The problems were interesting, though. Ilse was asked to chart the nonluminous bodies in the Solar System. She became quite skilled at this and eventually discovered all nine planets, most of their moons, and several asteroids and comets.

In less than two years, Ilse was farther from the sun than any known planet, than any previous terrestrial probe. The sun itself was no more than a very bright star behind her, and Ilse had no trouble keeping her frigid innards at their proper temperature. But now it took sixteen hours to ask a question of Earth and obtain an answer.

A strange thing happened. Over a period of three weeks, the sun became steadily brighter until it gleamed ten times as luminously as before. The change was not really a great one. It was far short of what Earth's astronomers would have called a nova. Nevertheless, Ilse puzzled over the event, in her own way, for many months, since it was at this time that she lost maser contact with Earth. That contact was never regained.

Now Ilse changed herself to meet the empty centuries. As her designers had planned, she split her mind into three coequal entities. Theoretically each of these minds could handle the entire mission alone, but for any important decision, Ilse required the agreement of at least two of the minds. In this fractionated state, Ilse was neither as bright nor as quick-thinking as she had been at launch. But scarcely anything happened in interstellar space, the chief danger being senile decay. Her three minds spent as much time checking one another as they did overseeing the various subsystems.

The one thing they did not regularly check was the programmed memory, since Ilse's designers had – mistakenly – judged that such checks were a greater danger to the memories than the passage of time.

Even with her mentality diminished, and in spite of the caretaker tasks assigned her, Ilse spent much of her time contemplating the universe that spread out forever around her. She discovered binary star systems, then watched the tiny lights swing back and forth around each other as the decades and centuries passed. To her the universe became a moving, almost a living, thing. Several of the nearer stars drifted almost a degree every century, while the great galaxy in Andromeda shifted less than a second of arc in a thousand years.

Occasionally, she turned about to look at Sol. Even ten centuries out she could still distinguish Jupiter and Saturn. These were auspicious observations.

Finally it was time for the mid-course correction. She had spent the preceding century refining her alignment and her navigational observations. The burn was to be only one hundred meters per second, so accurate had been her perihelion impulse. Nevertheless, without that correction she would miss the Centauran system entirely. When the second arrived and her alignment was perfect, Ilse lit her tiny rocket – and discovered that she could obtain at most only three quarters of the rated thrust. She had to make two burns before she was satisfied with the new course.

For the next fifty years, Ilse studied the problem. She tested the rocket's electrical system hundreds of times, and even fired the rocket in microsecond bursts. She never discovered how the centuries had robbed her, but extrapolating from her observations, Ilse realized that by the time she entered the Centauran system, she would have only a thousand meters per second left in her rocket – less than half its designed capability. Even so it was possible that, without further complications, she would be able to survey the planets of both stars in the system.

But before she finished her study of the propulsion problem, Ilse discovered another breakdown – the most serious she was to face:

She had forgotten her mission. Over the centuries the pattern of magnetic fields on her programmed memory had slowly disappeared – the least used programs going first. When Ilse recalled those programs to discover how her reduced maneuverability affected the mission, she discovered that she no longer had any record of her ultimate purpose. The memories ended with badly faded programs for biochemical reconnaissance and planetary entry, and Ilse guessed that there was something crucial left to do after a successful landing on a suitable planet.

Ilse was a patient sort – especially in her cruise configuration – and she didn't worry about her ultimate purpose, so far away in the future. But she did do her best to preserve what programs were left. She played each program into her own memory and then back to the programmed memory. If the process were repeated every seventy years, she found that she could keep the programmed memories from fading. On the other hand, she had no way of knowing how many errors this endless repetition was introducing. For this reason she had each of her subminds perform the process separately, and she frequently checked the ballistic and astronomical programs by doing problems with them.

Ilse went further: she studied her own body for clues as to its purpose. Much of her body was filled with a substance she must keep within a few degrees of absolute zero. Several leads disappeared into this mass. Except for her thermometers, however, she had no feeling in this part of her body. Now she raised the temperature in this section a few thousandths of a degree, a change well within design specifications, but large enough for her to sense. Comparing her observations and the section's mass with her chemical analysis programs, Ilse concluded that the mysterious area was a relatively homogeneous body of frozen water, doped with various impurities. It was interesting information, but no matter how she compared it with her memories she could not see any significance to it.

Ilse floated on – and on. The period of time between the midcourse maneuver and the next important event on her schedule was longer than Man's experience with agriculture had been on Earth.

As the centuries passed, the two closely set stars that were her destination became brighter until, a thousand years from Alpha Centauri, she decided to begin her search for planets in the system. Ilse turned her telescope on the brighter of the two stars ... call it Able. She was still thirty-five thousand times as far from Able – and the smaller star ... call it Baker – as Earth is from Sol. Even to her sharp eye, Able didn't show as a disk but rather as a diffraction pattern: a round blob of light – many times larger than the star's true disk – surrounded by a ring of light. The faint gleam of any planets would be lost in that diffraction pattern. For five years Ilse watched the pattern, analyzed it with one of her most subtle programs. Occasionally she slid occulting plates into the telescope and studied the resulting, distorted, pattern. After five years she had found suggestive anomalies in the diffraction pattern, but no definite signs of planets.

No matter. Patient Ilse turned her telescope a tiny fraction of a degree, and during the next five years she watched Baker. Then she switched back to Able. Fifteen times Ilse repeated this cycle. While she watched, Baker completed two revolutions about Able, and the stars' maximum mutual separation increased to nearly a tenth of a degree. Finally Ilse was certain: she had discovered a planet orbiting Baker, and perhaps another orbiting Able. Most likely they were both gas giants. No matter: she knew that any small, inner planets would still be lost in the glare of Able and Baker.

There remained less than nine hundred years before she coasted through the Centauran system.

Ilse persisted in her observations. Eventually she could see the gas giants as tiny spots of light – not merely as statistical correlations in her carefully collected diffraction data. Four hundred years out, she decided that the remaining anomalies in Able's diffraction pattern must be another planet, this one at about the same distance from Able as Earth is from Sol. Fifteen years later she made a similar discovery for Baker.

If she were to investigate both of these planets she would have to plan very carefully. According to her design specifications, she had scarcely the maneuvering capability left to investigate one system. But Ilse's navigation system had survived the centuries better than expected, and she estimated that a survey of both planets might still be possible.

Three hundred and fifty years out, Ilse made a relatively large course correction, better than two hundred meters per second. This change was essentially a matter of pacing: It would delay her arrival by four months. Thus she would pass near the planet she wished to investigate and, if no landing were attempted, her path would be precisely bent by Able's gravitational field and she would be cast into Baker's planetary system.

Now Ilse had less than eight hundred meters per second left in her rocket – less than one percent of her velocity relative to Able and Baker. If she could be at the right place at the right time, that would be enough, but otherwise ...

Ilse plotted the orbits of the bodies she had detected more and more accurately. Eventually she discovered several more planets: a total of three for Able, and four for Baker. But only her two prime candidates – call them Able II and Baker II – were at the proper distance from their suns.

Eighteen months out, Ilse sighted moons around Able II. This was good news. Now she could accurately determine the planet's mass, and so refine her course even more. Ilse was now less than fifty astronomical units from Able, and eighty from Baker. She had no trouble making spectroscopic observations of the planets. Her prime candidates had plenty of oxygen in their atmospheres – though the farther one, Baker II, seemed deficient in water vapor. On the other hand, Able II had complex carbon compounds in its atmosphere, and its net color was blue green. According to Ilse's damaged memory, these last were desirable features.

The centuries had shrunk to decades, then to years, and finally to days. Ilse was within the orbit of Able's gas giant. Ten million kilometers ahead her target swept along a nearly circular path about its sun, Able. Twenty-seven astronomical units beyond Able gleamed Baker.

But Ilse kept her attention on that target, Able II. Now she could make out its gross continental outlines. She selected a landing site, and performed a two hundred meter per second burn. If she chose to land, she would come down in a greenish, beclouded area.

Twelve hours to contact. Ilse checked each of her subminds one last time. She deleted all malfunctioning circuits, and reassembled herself as a single mind out of what remained. Over the centuries, one third of all her electrical components had failed, so that besides her lost memories, she was not nearly as bright as she had been when launched. Nevertheless, with her subminds combined she was much cleverer than she had been during the cruise. She needed this greater alertness, because in the hours and minutes preceding her encounter with Able II, she would do more analysis and make more decisions than ever before.

One hour to contact. Ilse was within the orbit of her target's outer moon. Ahead loomed the tentative destination, a blue and white crescent two degrees across. Her landing area was around the planet's horizon. No matter. The important task for these last moments was a biochemical survey – at least that's what her surviving programs told her. She scanned the crescent, looking for traces of green through the clouds. She found a large island in a Pacific-sized ocean, and began the exquisitely complex analysis necessary to determine the orientation of amino acids. Every fifth second, she took one second to reestimate the atmospheric densities. The problems seemed even more complicated than her training exercises back in Earth orbit.

Five minutes to contact. She was less than forty thousand kilometers out, and the planet's hazy limb filled her sky. In the next ten seconds she must decide whether or not to land on Able II. Her ten-thousand-year mission was at stake here. For once Ilse landed, she knew that she would never fly again. Without the immense booster that had pushed her out along this journey, she was nothing but a brain and an entry shield and a chunk of frozen water. If she decided to bypass Able II, she must now use a large portion of her remaining propellants to accelerate at right angles to her trajectory. This would cause her to miss the upper edge of the planet's atmosphere, and she would go hurtling out of Able's planetary system. Thirteen months later she would arrive in the vicinity of Baker, perhaps with enough left in her rocket to guide herself into Baker II's atmosphere. But, if that planet should be inhospitable, there would be no turning back: she would have to land there, or else coast on into interstellar darkness.

Ilse weighed the matter for three seconds and concluded that Able II satisfied every criterion she could recall, while Baker II seemed a bit too yellow, a bit too dry.

Ilse turned ninety degrees and jettisoned the small rocket that had given her so much trouble. At the same time she ejected the telescope which had served her so well. She floated indivisible, a white biconvex disk, twelve meters in diameter, fifteen tons in mass.

She turned ninety degrees more to look directly back along her trajectory. There was not much to see now that she had lost her scope, but she recognized the point of light that was Earth's sun and wondered again what had been on all those programs that she had forgotten.

Five seconds. Ilse closed her eye and waited.

Contact began as a barely perceptible acceleration. In less than two seconds that acceleration built to two hundred and fifty gravities. This was beyond Ilse's experience, but she was built to take it: her body contained no moving parts and – except for her fusion reactor – no empty spaces. The really difficult thing was to keep her body from turning edgewise and burning up. Though she didn't know it, Ilse was repeating – on a grand scale – the landing technique that men had used so long ago. But Ilse had to dissipate more than eight hundred times the kinetic energy of any returning Apollo capsule. Her maneuver was correspondingly more dangerous, but since her designers could not equip her with a rocket powerful enough to decelerate her, it was the only option.

Now Ilse used her wits and every dyne in her tiny electric thrusters to arc herself about Able II at the proper attitude and altitude. The acceleration rose steadily toward five hundred gravities, or almost five kilometers per second in velocity lost every second. Beyond that Ilse knew that she would lose consciousness. Just centimeters away from her body the air glowed at fifty thousand degrees. The fireball that surrounded her lit the ocean seventy kilometers below as with daylight.

Four hundred and fifty gravities. She felt a cryostat shatter, and one branch of her brain short through. Still Ilse worked patiently and blindly to keep her body properly oriented. If she had calculated correctly, there were less than five seconds to go now.

She came within sixty kilometers of the surface, then rose steadily back into space. But now her velocity was only seven kilometers per second. The acceleration fell to a mere fifteen gravities, then to zero. She coasted back through a long ellipse to plunge, almost gently, into the depths of Able II's atmosphere.

At twenty thousand meters altitude, Ilse opened her eye and scanned the world below. Her lens had been cracked, and several of her gestalt programs damaged, but she saw green and knew her navigation hadn't been too bad.

It would have been a triumphant moment if only she could have remembered what she was supposed to do ***after*** she landed.

At ten thousand meters, Ilse popped her paraglider from the hull behind her eye. The tough plastic blossomed out above her, and her fall became a shallow glide. Ilse saw that she was flying over a prairie spotted here and there by forest. It was near sunset and the long shadows cast by trees and hills made it easy for her to gauge the topography.

Two thousand meters. With a glide ratio of one to four, she couldn't expect to fly more than another eight kilometers. Ilse looked ahead, saw a tiny forest, and a stream glinting through the trees. Then she saw a glade just inside the forest, and some vagrant memory told her this was an appropriate spot. She pulled in the paraglider's forward lines and slid more steeply downwards. As she passed three or four meters over the trees surrounding the glade, Ilse pulled in the rear lines, stalled her glider, and fell into the deep, moist grass. Her dun and green paraglider collapsed over her charred body so that she might be mistaken for a large black boulder covered with vegetation.

The voyage that had crossed one hundred centuries and four light-years was ended.

Ilse sat in the gathering twilight and listened. Sound was an undreamed of dimension to her: tiny things burrowing in their holes, the stream gurgling nearby, a faint chirping in the distance. Twilight ended and a shallow fog rose in the dark glade. Ilse knew her voyaging was over. She would never more again. No matter. That had been planned, she was sure. She knew that much of her computing machinery–her mind–had been destroyed in the landing. She would not survive as a conscious being for more than another century or two. No matter.

What did matter was that she knew that her mission was not completed, and that the most important part remained, else the immense gamble her makers had undertaken would finally come to nothing. That possibility was the only thing which could frighten Ilse. It was part of her design.

She reviewed all the programmed memories that had survived the centuries and the planetary entry, but discovered nothing new. She investigated the rest of her body, testing her parts in a thorough, almost destructive, way she never would have dared while still centuries from her destination. She discovered nothing new. Finally she came to that load of ice she had carried so far. With one of her cryostats broken, she couldn't keep it at its proper temperature for more than a few years. She recalled the apparently useless leads that disappeared into that mass. There was only one thing left to try.

Ilse turned down her cryostats, and waited as the temperature within her climbed. The ice near her small fusion reactor warmed first. Somewhere in the frozen mass a tiny piece of metal expanded just far enough to complete a circuit, and Ilse discovered that her makers had taken one last precaution to insure her reliability. At the base of the icy hulk, next to the reactor, they had placed an auxiliary memory unit, and now Ilse had access to it. Her designers had realized that no matter what dangers they imagined, there would be others, and so they had decided to leave this back-up cold and inactive till the very end. And the new memory unit was quite different from her old ones, Ilse vaguely realized. It used optical rather than magnetic storage.

Now Ilse knew what she must do. She warmed a cylindrical tank filled with frozen amniotic fluid to thirty-seven degrees centigrade. From the store next to the cylinder, she injected a single microorganism into the tank. In a few minutes she would begin to suffuse blood through the tank.

It was early morning now and the darkness was moist and cool. Ilse tried to probe her new memory further, but was balked. Apparently the instructions were delivered according to some schedule to avoid unnecessary use of the memory. Ilse reviewed what she had learned, and decided that she would know more in another nine months.

After You Have Read the Story …

The final resolution of Ilse's mission packs the emotional wallop of Theodore Sturgeon's sci-fi short story classic "The Man Who Lost the Sea," echoing the "God, we made it!" ending.[6] The twist of the mission objective revealed on landing was clearly intended to cause a lump in the throat, but in terms of artificial intelligence, just reaching the planet is more than sufficient cause for celebration and admiration for the intrepid Ilse and her long-dead designers.

"Long Shot" illustrates why AI for robotics usually makes an open world assumption and how a robot might use each of the four categories of deliberation. These four categories, generating, selecting, implementing, and monitoring, are intuitive components of making a plan and following it through. Often, though, robotics concentrates on generating, selecting, and implementing a plan while forgetting to explicitly monitor for problems. Ilse, however, uses all four categories. The problems she encounters while keeping to the nominal mission schedule illustrate how different types of deliberation happen over different time horizons, which can be exploited to produce elegant, distributed, asynchronous computation.

Open versus Closed World Assumptions

Reactive behaviors such as those described in "Runaround" allow robots to function in an open world because they don't depend on accurate knowledge representations and planning. A robot doesn't have to recognize that a chair is in its way and plan an optimal path around it, only that something is in its way, and then be repulsed by it. Thus if a robot encounters something it has never seen before, for example that there is literally an elephant in the room, it can still go around it.

Although reactive behaviors let a robot function in the open world, those behaviors don't produce a robot much smarter than a cockroach or a fish. For a spaceship like Ilse, more sophisticated intelligence is in order. Ilse must be able to deliberate on her state of existence, repair herself as needed, adjust her course, and plan for contingencies. In other words, she needs a *global world model*, a representation of knowledge about the world. A portion of the representation might be a three-dimensional reconstruction of the world for navigation and spatial reasoning, but the world model would likely also contain semantic labels of objects (e.g., "that is a coffee cup" and "this is my coffee cup that was given to me on my thirty-third birthday"). Deliberation in the open world is challenging because of the paradox of representing, and labeling, the unknown.

Four Categories of Deliberation

Autonomous robots are generally built to have a basic layer of reactive behaviors along with a deliberative layer handling the more sophisticated components of intelligence. Behaviors don't need a world model because they react directly to what is being sensed. Deliberative functions have to work with a broad set of knowledge, from the layout of a room to what is intended by a command. As previously mentioned, deliberative components fall into four categories: generating a plan or solving a problem (which are often identical algorithmically), selecting the resources, such as sensors and actuators, by which to accomplish the plan, implementing the plan by turning on behaviors, and then monitoring the execution of the plan. Figure 3.1 shows the four categories graphically anchored by a global world model forming a deliberative layer that sits on top of the reactive layer.

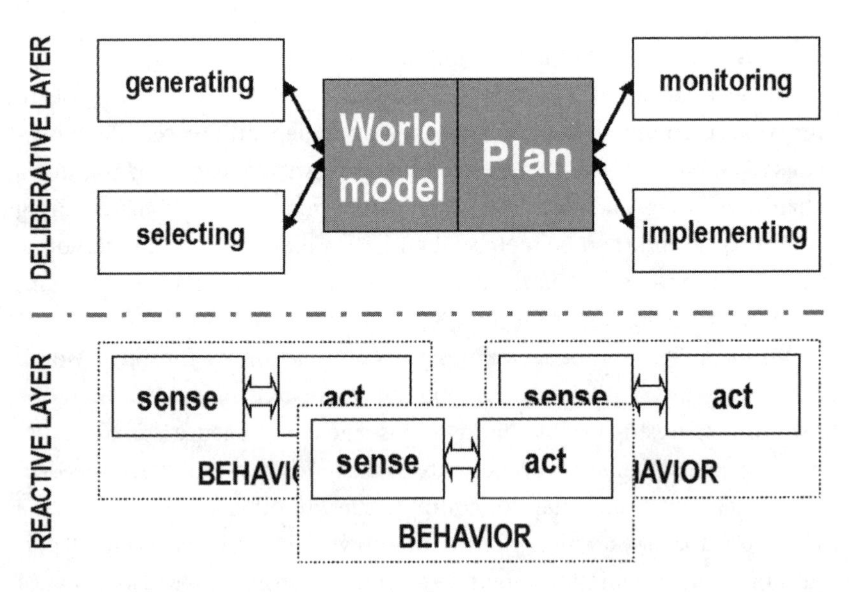

Figure 3.1
Deliberative components in a robot architecture.

"Long Shot" exemplifies the generating, selecting, and monitoring deliberative activities. Ilse first starts generating plans for navigation, such as ballistic trajectories, and even overrides human instruction on when to commence a burn. These are examples of planning via search techniques, where it is a matter of managing extreme computations to find an optimal answer.

But Ilse also performs planning that requires inference. Search in AI refers to searching for a "right" answer to be found by exploring a computational space; the answer is explicitly in the computational space somewhere. Inference in AI means that the answer isn't directly there—there are missing links or data—and thus the inference algorithm has to fill in those gaps. Choosing Able II requires forging implicit connections between the data Ilse has gathered and what desirable features are so that she can eventually make a judgment call. The story does not specifically comment on how Ilse infers Able II is the best choice, but it reads as if it is a type of inference known as analogical or case-based reasoning. Able II matches all of the important features of Earth.

Inferring by making comparisons is not the only type of inference. In recent years, research in inference algorithms, especially for data mining, has begun to rely more heavily on probability theory to make missing connections in a knowledge base. Logical inference is another style of inference.

The actual mechanism for selecting the right actions to perform is also touched on in the story. Like any well-engineered AI robotics architecture, Ilse has a library of programs that appear to be the same as the AI knowledge representation called *scripts*, and they enable various capabilities and sequences of functions. Scripts are similar to finite state machines but favor semantic shortcuts. "Long Shot" comments on how Ilse is noteworthy because she can select the more appropriate script with incomplete information; this ability to work in the presence of incomplete information is another hallmark of inference.

Like Sturgeon's man who lost the sea but made a gigantic step for mankind, Ilse is heroic in persevering. In order to persist over the centuries, Ilse appears to use the classic General Problem Solver paradigm in planning initially developed by Allen Newell and Herb Simon for Shakey, the first AI robot, built in 1967.[7] A strong point of Ilse's strategy is that she sometimes cannot remember the entire goal of the mission or how to accomplish it, but works on achieving the biggest subgoal that will get her closer to the overall goal. This is the heart of the *means-end analysis* technique in the General Problem Solver: if you can't plan a way that gets you to the goal, then plan to get as near to that goal as possible, with the expectation that once you have accomplished that subgoal things will have changed or new knowledge will have been added to the world model and at that point you can plan a way to reach the larger goal.

Ilse also monitors for the inevitable malfunctions that trigger her problem-solving capabilities to either fix or work around the problems. A significant failure for Ilse is the thruster rocket; in an instance of life imitating art, Deep Space One had a thruster valve that was stuck closed and its model-based reasoning system compensated by switching to a secondary control mode. Deep Space One also encountered and recovered from numerous failures, such as replanning to work around a stuck camera (reminiscent of one of the problems in *2001: A Space Odyssey*) and repairing an instrument by resetting it (also something Ilse does a lot of).[8]

Deliberation and Time Horizons

Another way of thinking about deliberation is in terms of time horizons. Reactive behaviors don't require deliberation because they only use information from the *present*; they react in stimulus-response fashion. Generating and monitoring plans requires a robot to understand what it should do in the *future* and what it has done in the *past* as well as what it is doing in the present. Once a plan is created, the selection and implementation activities may consider the past and present so as to keep track of plan execution, but normally those activities don't have to account for the future because the generating and monitoring functions are responsible for thinking ahead. Figure 3.2 shows the four categories of deliberation with their associated timelines.

Why should we care about time horizons in deliberation? One reason is that they impact the knowledge representations required to support deliberation. Deliberation can require storing information about the past, such as building a map or locating a robot in a sequence of events. It can also require projecting information about the future, which again means additional information must be stored. A second reason to care about time horizons is that they support distributed, asynchronous processing. Reactive behaviors can run on embedded processors at high update rates while

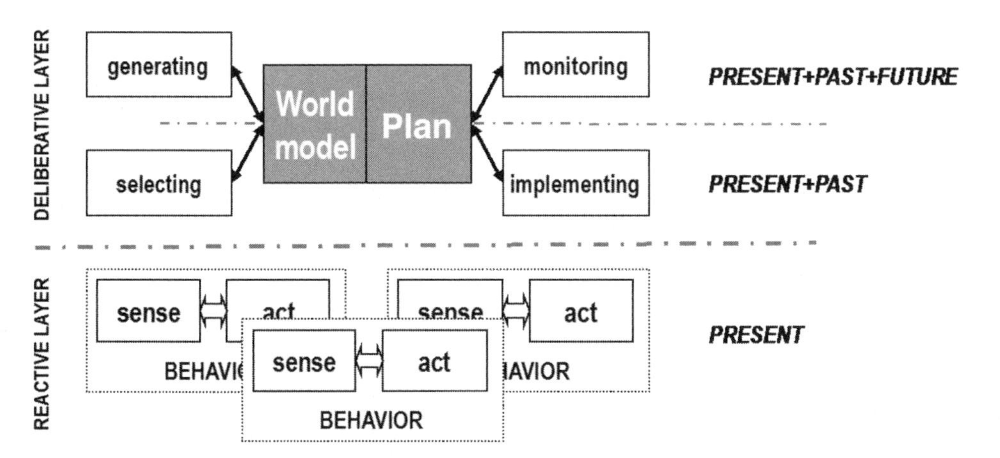

Figure 3.2
Time horizons in reaction and deliberation.

deliberative functions can simultaneously run on a different processor at lower update rates.

One of the earliest and most influential deliberative architectures was the NIST Real-Time Control Architecture.[9] It advocated dividing the set of functions into groups that ran independently every 50 milliseconds, 500 milliseconds, 1 second, and up to 10 minutes. Most AI roboticists use a hybrid deliberative/reactive architecture, sometimes called a three-layer architecture, rather than a purely deliberative one, but the hybrid architectures implicitly preserve the notion that reactive functions must react quickly to present stimulus while deliberative functions can take more time to compute and thus can be ported to distributed processors. Decomposing functions and implementing on separate processors can increase resilience to damage because if one processor fails, the others can still continue. The best example of this is in the book and movie *2001: A Space Odyssey*, when the modules containing HAL's higher deliberative functions that had turned murderous were yanked out but the functions needed to keep the ship running were left in. Ilse's subminds in "Long Shot" are a nod to the practicality of distributed, asynchronous processing.

Obviously "Long Shot" is about a time horizon ultimately spanning centuries. But time horizons are a concept central to deliberative intelligence and one that helps distinguish silicon-based agency (robots) from carbon-based agency (humans). The story is about how a robot ship might function for literally ten thousand years, making split-second decisions on spaceship control when a rocket isn't performing correctly yet at the same time anticipating and preparing for what might occur when using that rocket hundreds of years in the future. Ilse's ability to think and project over different time horizons means that the human race in the story will exist in the future long after the sun has gone nova. Perhaps our ability to imagine robots like Ilse means that we really are those humans.

Reality Score: A+

"Long Shot" gets it right; it is an accurate and plausible depiction of how an intelligent robot would work.

4 Testing: "Catch That Rabbit"

"Catch That Rabbit," written two years after "Runaround" in 1944, is another Three Laws of Robotics story featuring Mike Donovan and Greg Powell as consummate robot troubleshooters.[1] Greg and Mike appear in Asimov's stories as the robotics equivalent of Tom and Ray Magliozzi from NPR's "Car Talk" show, providing comic relief, whereas Dr. Susan Calvin, the icy robo-psychologist introduced in later *I, Robot* stories, is reserved for more cerebral mysteries. The Donovan and Powell stories are technology mystery stories that follow a basic pattern: one or more robots fail in some surprising way, there is something preventing Donovan and Powell from directly examining the system and figuring out the cause, and the rest of the story is spent with the two reasoning, and bickering, their way to the explanation.

In this story, DV-5, a robot that controls six other mining robots, inexplicably causes the team of robots to behave in odd and silly ways. Despite the team of seven robots, the technological MacGuffin in "Catch That Rabbit" is how to test that the system works and debug any problems, not distributed, multirobot control. Testing and debugging is a major challenge in artificial intelligence for robotics. Testing is important for software verification and validation (V&V), in this case so that US Robots and Mechanical Men can get paid. In software engineering, *verification* means the software is shown to meet the specifications and *validation* means the software is shown to meet the intent of the system. This distinction arose in software engineering because the intended functionality may be more than the sum of the individual component specifications. In "Catch That Rabbit," the DV-5 robot passes verification tests for controlling multiple robots but not validation tests.

As You Read the Story ...

Testing artificial intelligence is difficult because intelligence is nondeterministic. In computer science, algorithms are classified as either *deterministic* or *nondeterministic*. Deterministic algorithms take a set of inputs and produce only one possible output. These are algorithms for computing math problems, finite state machines, and optimizing planning and scheduling. They are best used in applications where there are discrete, clear cut values of inputs and states of outputs. Nondeterministic algorithms take a set of inputs and produce one of many possible outputs. The actual output may depend on timing between processes in the robot; if one process, such as a sensor input, finishes before another sensor updates, the data used by a motor control process may be slightly different and thus produce a slightly different action. Outputs may depend on sensor and hardware noise, which will change over time with wear and tear. The result can also depend on the sheer number of inputs, as it can become challenging to model the dependencies and interactions that lead to a particular output. The more sophisticated the robot, the more nondeterministic it is.

As you read, think about how to test the robots and specifically what could be tested deterministically and what would have to be tested nondeterministically. Speculate on what data or status information the robots could have provided that would have helped Donovan and Powell determine what was happening. Finally, consider whether you would have divided up the responsibilities of the robots the same way if they had been a team of human miners.

"Catch That Rabbit" by Isaac Asimov, 1944

The vacation was longer than two weeks, that, Mike Donovan had to admit. It had been six months, with pay. He admitted that, too. But that, as he explained furiously, was fortuitous. U. S. Robots had to get the bugs out of the multiple robots, and there were plenty of bugs, and there are always at least half a dozen bugs left for the field-testing. So they waited and relaxed until the drawing-board men and the slide-rule boys had said "OK!" And now he and Powell were out on the asteroid and it was not OK. He repeated that a dozen times, with a face that had gone beety, "For the love of Pete, Greg, get realistic. What's the use of adhering to the letter of the specifications and watching the test go to pot? It's about time you got the red tape out of your pants and went to work."

"I'm only saying," said Gregory Powell, patiently, as one explaining electronics to an idiot child, "that according to spec, those robots are equipped for asteroid mining without supervision. We're not supposed to watch them."

"All right. Look – logic!" He lifted his hairy fingers and pointed. "One: That new robot passed every test in the home laboratories. Two: United States Robots guaranteed their passing the test of actual performance on an asteroid. Three: The robots are not passing said tests. Four: If they don't pass, United States Robots loses ten million credits in cash and about one hundred million in reputation. Five: If they don't pass and we can't explain why they don't pass, it is just possible two good jobs may have to be bidden a fond farewell."

Powell groaned heavy behind a noticeably insincere smile. The unwritten motto of United States Robot and Mechanical Men Corp. was well known: "No employee makes the same mistake twice. He is fired the first time."

Aloud he said, "You're as lucid as Euclid with everything except the facts. You've watched that robot group for three shifts, you redhead, and they did their work perfectly. You said so yourself. What else can we do?"

"Find out what's wrong, that's what we can do. So they did work perfectly when I watched them. But on three different occasions when I didn't watch them, they didn't bring in any ore. They didn't even come back on schedule. I had to go after them."

"And was anything wrong?"

"Not a thing. Not a thing. Everything was perfect. Smooth and perfect as the luminiferous ether. Only one little insignificant detail disturbed me – there was no ore."

Powell scowled at the ceiling and pulled at his brown mustache. "I'll tell you what, Mike. We've been stuck with pretty lousy jobs in our time, but this takes the iridium asteroid. The whole business is complicated past endurance. Look, that robot, DV-5, has six robots under it. And not just under it – they're part of it."

"I know that – "

"Shut up!" said Powell, savagely, "I know you know it, but I'm just describing the hell of it. Those six subsidiaries are part of DV-5 like your fingers are part of you and it gives them their orders neither by voice nor radio, but directly through positronic fields. Now – there isn't a roboticist back at United States Robots that knows what a positronic field is or how it works. And neither do I. Neither do you."

"The last," agreed Donovan, philosophically, "I know."

"Then look at our position. If everything works – fine! If anything goes wrong – we're out of our depth and there probably isn't a thing we can do, or anybody else. But the job belongs to us and not to anyone else so we're on the spot, Mike." He blazed away for a moment in silence. Then, "All right, have you got him outside?"

"Yes."

"Is everything normal now?"

"Well he hasn't got religious mania, and he isn't running around in a circle spouting Gilbert and Sullivan, so I suppose he's normal."

Donovan passed out the door, shaking his head viciously.

Powell reached for the "Handbook of Robotics" that weighed down one side of his desk to a near-founder and opened it reverently. He had once jumped out of the window of a burning house dressed only in shorts and the "Handbook." In a pinch, he would have skipped the shorts.

The "Handbook" was propped up before him, when Robot DV-5 entered, with Donovan kicking the door shut behind him.

Powell said somberly, "Hi, Dave. How do you feel?"

"Fine," said the robot. "Mind if I sit down?" He dragged up the specially reinforced chair that was his, and folded gently into it.

Powell regarded Dave – laymen might think of robots by their serial numbers; roboticists never – with approval. It was not overmassive by any means, in spite of its construction as thinking-unit of an integrated seven-unit robot team. It was seven feet tall, and a halfton of metal and electricity. A lot? Not when that half-ton has to be a mass of condensers, circuits, relays, and vacuum cells that can handle practically any psychological reaction known to humans. And a positronic brain, which with ten pounds of matter and a few quintillions of positrons runs the whole show.

Powell groped in his shirt pocket for a loose cigarette. "Dave," he said, "you're a good fellow. There's nothing flighty or prima donnaish about you. You're a stable, rockbottom mining robot, except that you're equipped to handle six subsidiaries in direct coordination. As far as I know, that has not introduced any unstable paths in your brain-path map."

The robot nodded, "That makes me feel swell, but what are you getting at, boss?" He was equipped with an excellent diaphragm, and the presence of overtones in the sound unit robbed him of much of that metallic flatness that marks the usual robot voice.

"I'm going to tell you. With all that in your favor, what's going wrong with your job? For instance, today's B-shift?"

Dave hesitated, "As far as I know, nothing."

"You didn't produce any ore."

"I know."

"Well, then – "

Dave was having trouble, "I can't explain that, boss. It's been giving me a case of nerves, or it would if I let it – my subsidiaries worked smoothly. I know I did." He considered, his photoelectric eyes glowing intensely. Then, "I don't remember. The day ended and there was Mike and there were the ore cars, mostly empty."

Donovan broke in, "You didn't report at shift-end those days, Dave. You know that?"

"I know. But as to why – " He shook his head slowly and ponderously.

Powell had the queasy feeling that if the robot's face were capable of expression, it would be one of pain and mortification. A robot, by its very nature, cannot bear to fail its function.

Donovan dragged his chair up to Powell's desk and leaned over, "Amnesia, do you think?"

"Can't say. But there's no use in trying to pin disease names on this. Human disorders apply to robots only as romantic analogies. They're no help to robotic engineering." He scratched his neck; "I hate to put him through the elementary brain-reaction tests. It won't help his self-respect any."

He looked at Dave thoughtfully and then at the Field-Test outline given in the "Handbook." He said, "See here, Dave, what about sitting through a test? It would be the wise thing to do."

The robot rose, "If you say so, boss." There was pain in his voice.

It started simply enough. Robot DV-5 multiplied five-place figures to the heartless ticking of a stopwatch. He recited the prime numbers between a thousand and ten thousand. He extracted cube roots and integrated functions of varying complexity. He went through mechanical reactions in order of increasing difficulty. And, finally, worked his precise mechanical mind over the highest function of the robot world – the solutions of problems in judgment and ethics.

At the end of two hours, Powell was copiously besweated. Donovan had enjoyed a none-too-nutritious diet of fingernail and the robot said, "How does it look, boss?"

Powell said, "I've got to think it over, Dave. Snap judgments won't help much. Suppose you go back to the C-shift. Take it easy. Don't press too hard for quota just for a while – and we'll fix things up."

The robot left. Donovan looked at Powell.

"Well – "

Powell seemed determined to push up his mustache by the roots. He said, "There is nothing wrong with the currents of his positronic brain."

"I'd hate to be that certain."

"Oh, Jupiter, Mike! The brain is the surest part of a robot. It's quintuple-checked back on Earth. If they pass the field test perfectly, the way Dave did, there just isn't a chance of brain misfunction. That test covered every key path in the brain."

"So where are we?"

"Don't rush me. Let me work this out. There's still the possibility of a mechanical breakdown in the body. That leaves about fifteen hundred condensers, twenty thousand individual electric circuits, five hundred vacuum cells, a thousand relays, and upty-ump thousand other individual pieces of complexity that can be wrong. And these mysterious positron is fields no one knows anything about."

"Listen, Greg," Donovan grew desperately urgent. "I've got an idea. That robot may be lying. He never – "

"Robots can't knowingly lie, you fool. Now if we had the McCormack-Wesley tester, we could check each individual item in his body within twenty-four to forty-eight hours, but the only two M-W testers existing are on Earth, and they weigh ten tons, are on concrete foundations, and can't be moved. Isn't that peachy?"

Donovan pounded the desk, "But, Greg, he only goes wrong when we're not around. There's something – sinister – about – that." He punctuated the sentence with slams of fist against desk.

"You," said Powell, slowly, "make me sick. You've been reading adventure novels."

"What I want to know," shouted Donovan, "is what we're going to do about it."

"I'll tell you. I'm going to install a visiplate right over my desk. Right on the wall over there, see!" He jabbed a vicious finger at the spot. "Then I'm going to focus it at whatever part of the mine is being worked, and I'm going to watch. That's all."

"That's all? Greg – "

Powell rose from his chair and leaned his balled fists on the desk, "Mike, I'm having a hard time." His voice was weary. "For a week, you've been plaguing me about Dave. You say he's gone wrong. Do you know how he's gone wrong? No! Do you know what shape this wrongness takes? No! Do you know what brings it on? No! Do you know what snaps him out? No! Do you know anything about it? No! Do I know anything about it? No! So what do you want me to do?"

Donovan's arm swept outward in a vague, grandiose gesture, "You got me!"

"So I tell you again. Before we do anything toward a cure, we've got to find out what the disease is in the first place. The first step in cooking rabbit stew is catching the rabbit. Well, we've got to catch that rabbit! Now get out of here."

Donovan stared at the preliminary outline of his field report with weary eyes. For one thing, he was tired and for another, what was there to report while things were unsettled? He felt resentful.

He said, "Greg, we're almost a thousand tons behind schedule."

"You," replied Powell, never looking up, "are telling me something I don't know."

"What I want to know," said Donovan, in sudden savagery, "is why we're always tangled up with new-type robots. I've finally decided that the robots that were good enough for my great-uncle on my mother's side are good enough for me. I'm for what's tried and true. The test of time is what counts – good, solid, old-fashioned robots that never go wrong."

Powell threw a book with perfect aim, and Donovan went tumbling off his seat.

"Your job," said Powell, evenly, "for the last five years has been to test new robots under actual working conditions for United States Robots. Because you and I have been so injudicious as to display proficiency at the task, we've been rewarded with the dirtiest jobs. That," he jabbed holes in the air with his finger in Donovan's direction, "is your work. You've been griping about it, from personal memory, since about five minutes after United States Robots signed you up. Why don't you resign?"

"Well, I'll tell you." Donovan rolled onto his stomach, and took a firm grip on his wild, red hair to hold his head up. "There's a certain principle involved. After all, as a troubleshooter, I've played a part in the development of new robots. There's the principle of aiding scientific advance. But don't get me wrong. It's not the principle that keeps me going; it's the money they pay us. Greg!"

Powell jumped at Donovan's wild shout, and his eyes followed the red-head's to the visiplate, when they goggled in fixed horror. He whispered, "Holy – howling – Jupiter!"

Donovan scrambled breathlessly to his feet, "Look at them, Greg. They've gone nuts."

Powell said, "Get a pair of suits. We're going out there."

He watched the posturings of the robots on the visiplate. They were bronzy gleams of smooth motion against the shadowy crags of the airless asteroid. There was a marching formation now, and in their own dim body light, the roughhewn walls of the mine tunnel swam past noiselessly, checkered with misty erratic blobs of shadow. They marched in unison, seven of them, with Dave at the head. They wheeled and turned in macabre simultaneity; and melted through changes of formation with the weird ease of chorus dancers in Lunar Bowl.

Donovan was back with the suits, "They've gone jingo on us, Greg. That's a military march."

"For all you know," was the cold response, "it may be a series of callisthenic exercises. Or Dave may be under the hallucination of being a dancing master. Just you think first, and don't bother to speak afterward, either."

Donovan scowled and slipped a detonator into the empty side holster with an ostentatious shove. He said, "Anyway, there you are. So we work with new-model robots. It's our job, granted. But answer me one question. Why ... why does something invariably go wrong with them?"

"Because," said Powell, somberly, "we are accursed. Let's go!"

Far ahead through the thick velvety blackness of the corridors that reached past the illuminated circles of their flashlights, robot light twinkled.

"There they are," breathed Donovan.

Powell whispered tensely, "I've been trying to get him by radio but he doesn't answer. The radio circuit is probably out."

"Then I'm glad the designers haven't worked out robots who can work in total darkness yet. I'd hate to have to find seven mad robots in a black pit without radio communication, if they weren't lit up like blasted radioactive Christmas trees."

"Crawl up on the ledge above, Mike. They're coming this way, and I want to watch them at close range. Can you make it?"

Donovan made the jump with a grunt. Gravity was considerably below Earth-normal, but with a heavy suit, the advantage was not too great, and the ledge meant a near ten-foot jump. Powell followed.

The column of robots was trailing Dave single-file. In mechanical rhythm, they converted to double and returned to single in different order. It was repeated over and over again and Dave never turned his head.

Dave was within twenty feet when the play-acting ceased. The subsidiary robots broke formation, waited a moment, then clattered off into the distance – very rapidly. Dave looked after them, then slowly sat down. He rested his head in one hand in a very human gesture.

His voice sounded in Powell's earphones, "Are you here, boss?"

Powell beckoned to Donovan and hopped off the ledge.

"OK, Dave, what's been going on?"

The robot shook his head, "I don't know. One moment I was handling a tough outcropping in Tunnel 17, and the next I was aware of humans close by, and I found myself half a mile down main-stem."

"Where are the subsidiaries now?" asked Donovan.

"Back at work, of course. How much time has been lost?"

"Not much. Forget it." Then to Donovan, Powell added, "Stay with him the rest of the shift. Then, come back. I've got a couple of ideas."

It was three hours before Donovan returned. He looked tired. Powell said, "How did it go?"

Donovan shrugged wearily, "Nothing ever goes wrong when you watch them. Throw me a butt, will you?"

The redhead lit it with exaggerated care and blew a careful smoke ring. He said, "I've been working it out, Greg. You know, Dave has a queer background for a robot. There are six others under him in an extreme regimentation. He's got life and death power over those subsidiary robots and it must react on his mentality. Suppose he finds it necessary to emphasize this power as a concession to his ego."

"Get to the point."

"It's right here. Suppose we have militarism. Suppose he's fashioning himself an army. Suppose – he's training them in military maneuvers. Suppose – "

"Suppose you go soak your head. Your nightmares must be in technicolor. You're postulating a major aberration of the positronic brain. If your analysis were correct, Dave would have to break down the First Law of Robotics: that a robot may not injure a human being or, through inaction, allow a human being to be injured. The type of militaristic attitude and domineering ego you propose must have as the end-point of its logical implications, domination of humans."

"All right. How do you know that isn't the fact of the matter?"

"Because any robot with a brain like that would, one, never have left the factory, and two, be spotted immediately if it ever was. I tested Dave, you know."

Powell shoved his chair back and put his feet on the desk. "No. We're still in the position where we can't make our stew because we haven't the slightest notion as to what's wrong. For instance, if we could find out what that dance macabre we witnessed was all about, we would be on the way out."

He paused, "Now listen, Mike, how does this sound to you? Dave goes wrong only when neither of us is present. And when he is wrong, the arrival of either of us snaps him out of it."

"I once told you that was sinister."

"Don't interrupt. How is a robot different when humans are not present? The answer is obvious. There is a larger requirement of personal initiative. In that case, look for the body parts that are affected by the new requirements."

"Golly." Donovan sat up straight, then subsided. "No, no. Not enough. It's too broad. It doesn't cut the possibilities much."

"Can't help that. In any case, there's no danger of not making quota. We'll take shifts watching those robots through the visor. Any time anything goes wrong, we get to the scene of action immediately. That will put them right."

"But the robots will fail spec anyway, Greg. United States Robots can't market DV models with a report like that."

"Obviously. We've got to locate the error in make-up and correct it – and we've got ten days to do it in." Powell scratched his head. "The trouble is ... well, you had better look at the blueprints yourself."

The blueprints covered the floor like a carpet and Donovan crawled over the face of them following Powell's erratic pencil.

Powell said, "Here's where you come in, Mike. You're the body specialist, and I want you to check me. I've been trying to cut out all circuits not involved in the personal initiative hookup. Right here, for instance, is the trunk artery involving mechanical operations. I cut out all routine side routes as emergency divisions – " He looked up, "What do you think?"

Donovan had a very bad taste in his mouth, "The job's not that simple, Greg. Personal initiative isn't an electric circuit you can separate from the rest and study. When a robot is on his own, the intensity of the body activity increases immediately on almost all fronts. There isn't a circuit entirely unaffected. What must be done is to locate the particular condition – a very specific condition – that throws him off, and then start eliminating circuits."

Powell got up and dusted himself, "Hmph. All right. Take away the blueprints and burn them."

Donovan said, "You see when activity intensifies, anything can happen, given one single faulty part. Insulation breaks down, a condenser spills over, a connection sparks, a coil overheats. And if you work blind, with the whole robot to choose from, you'll never find the bad spot. If you take Dave apart and test every point of his body mechanism one by one, putting him together each time, and trying him out."

"All right. All right. I can see through a porthole, too."

They faced each other hopelessly, and then Powell said cautiously, "Suppose we interview one of the subsidiaries."

Neither Powell nor Donovan had ever had previous occasion to talk to a "finger." It could talk; it wasn't quite the perfect analogy to a human finger. In fact, it had a fairly developed brain, but that brain was tuned primarily to the reception of orders via positronic field, and its reaction to independent stimuli was rather fumbling.

Nor was Powell certain as to its name. Its serial number was DV-5-2, but that was not very useful.

He compromised. "Look, pal," he said, "I'm going to ask you to do some hard thinking and then you can go back to your boss."

The "finger" nodded its head stiffly, but did not exert its limited brainpower on speech.

"Now on four occasions recently," Powell said, "your boss deviated from brain-scheme. Do you remember those occasions?"

"Yes, sir."

Donovan growled angrily, "He remembers. I tell you there is something very sinister – "

"Oh, go bash your skull. Of course, the 'finger' remembers. There is nothing wrong with him." Powell turned back to the robot, "What were you doing each time ... I mean the whole group."

The "finger" had a curious air of reciting by rote, as if he answered questions by the mechanical pressure of his brainpan, but without any enthusiasm whatever.

He said, "The first time we were at work on a difficult outcropping in Tunnel 17, Level B. The second time we were buttressing the roof against a possible cave-in. The third time we were preparing accurate blasts in order to tunnel farther without breaking into a subterranean fissure. The fourth time was just after a minor cave-in."

"What happened at these times?"

"It is difficult to describe. An order would be issued, but before we could receive and interpret it, a new order came to march in queer formation."

Powell snapped out, "Why?"

"I don't know."

Donovan broke in tensely, "What was the first order ... the one that was superseded by the marching directions?"

"I don't know. I sensed that an order was sent, but there was never time to receive it."

"Could you tell us anything about it? Was it the same order each time?"

The "finger" shook his head unhappily, "I don't know."

Powell leaned back, "All right, get back to your boss."

The "finger" left, with visible relief.

Donovan said, "Well, we accomplished a lot that time. That was real sharp dialogue all the way through. Listen, Dave and that imbecile 'finger' are both holding out on us. There is too much they don't know and don't remember. We've got to stop trusting them, Greg."

Powell brushed his mustache the wrong way, "So help me, Mike, another fool remark out of you, and I'll take away your rattle and teething ring."

"All right. You're the genius of the team. I'm just a poor sucker. Where do we stand?"

"Right behind the eight ball. I tried to work it backward through the 'finger,' and couldn't. So we've got to work it forward."

"A great man," marveled Donovan. "How simple that makes it. Now translate that into English, Master."

"Translating it into baby talk would suit you better. I mean that we've got to find out what order it is that Dave gives just before everything goes black. It would be the key to the business."

"And how do you expect to do that? We can't get close to him because nothing will go wrong as long as we are there. We can't catch the orders by radio because they are transmitted via this positronic field. That eliminates the close-range and the long-range method, leaving us a neat, cozy zero."

"By direct observation, yes. There's still deduction."

"Huh?"

"We're going on shifts, Mike." Powell smiled grimly. "And we are not taking our eyes off the visiplate. We're going to watch every action of those steel headaches. When they go off into their act, we're going to see what happened immediately before and we're going to deduce the order."

Donovan opened his mouth and left it that way for a full minute. Then he said in strangled tones, "I resign. I quit."

"You have ten days to think up something better," said Powell wearily.

Which, for eight days, Donovan tried mightily to do. For eight days, on alternate four-hour shifts, he watched with aching and bleary eyes those glinty metallic forms move against the vague background. And for eight days in the four-hour in-betweens, he cursed United States Robots, the DV models, and the day he was born.

And then on the eighth day, when Powell entered with an aching head and sleepy eyes for his shift, Donovan stood up and with very careful and deliberate aim launched a heavy bookend for the exact center of the visiplate. There was a very appropriate splintering noise.

Powell gasped, "What did you do that for?"

"Because," said Donovan, almost calmly, "I'm not watching it any more. We've got two days left and we haven't found out a thing. DV-5 is a lousy loss. He's stopped five times since I've been watching and three times on your shift, and I can't make out what orders he gave, and you couldn't make it out. And I don't believe you could ever make it out because I know I couldn't ever."

"Jumping Space, how can you watch six robots at the same time? One makes with the hands, and one with the feet and one like a windmill and another is jumping up and down like a maniac. And the other two ... devil knows what they are doing. And then they all stop. So! So!"

"Greg, we're not doing it right. We got to get up close. We've got to watch what they're doing from where we can see the details."

Powell broke a bitter silence. "Yeah, and wait for something to go wrong with only two days to go."

"Is it any better watching from here?"

"It's more comfortable."

"Ah – But there's something you can do there that you can't do here."

"What's that?"

"You can make them stop – at whatever time you choose and while you're prepared and watching to see what goes wrong."

Powell startled into alertness, "Howzzat?"

"Well, figure it out, yourself. You're the brains you say. Ask yourself some questions. When does DV-5 go out of whack? When did that 'finger' say he did? When a cave-in threatened, or actually occurred, when delicately measured explosives were being laid down, when a difficult seam was hit."

"In other words, during emergencies," Powell was excited.

"Right! When did you expect it to happen! It's the personal initiative factor that's giving us the trouble. And it's just during emergencies in the absence of a human being that personal initiative is most strained. Now what is the logical deduction? How can we create our own stoppage when and where we want it?" He paused triumphantly – he was beginning to enjoy his role – and answered his own question to forestall the obvious answer on Powell's tongue. "By creating our own emergency."

Powell said, "Mike – you're right."

"Thanks, pal. I knew I'd do it some day."

"All right, and skip the sarcasm. We'll save it for Earth, and preserve it in jars for future long, cold winters. Meanwhile, what emergency can we create?"

"We could flood the mines, if this weren't an airless asteroid."

"A witticism, no doubt," said Powell. "Really, Mike, you'll incapacitate me with laughter. What about a mild cave-in?"

Donovan pursed his lips and said, "OK by me."

"Good. Let's get started."

Powell felt uncommonly like a conspirator as he wound his way over the craggy landscape. His sub-gravity walk teetered across the broken ground, kicking rocks to right and left under his weight in noiseless puffs of gray dust. Mentally, though, it was the cautious crawl of the plotter.

He said, "Do you know where they are?"

"I think so, Greg."

"All right," Powell said gloomily, "but if any 'finger' gets within twenty feet of us, we'll be sensed whether we are in the line of sight or not. I hope you know that."

"When I need an elementary course in robotics, I'll file an application with you formally, and in triplicate. Down through here."

They were in the tunnels now; even the starlight was gone. The two hugged the walls, flashes flickering out the way in intermittent bursts. Powell felt for the security of his detonator.

"Do you know this tunnel, Mike?"

"Not so good. It's a new one. I think I can make it out from what I saw in the visiplate, though – "

Interminable minutes passed, and then Mike said, "Feel that!"

There was a slight vibration thrumming the wall against the fingers of Powell's metal-incased hand. There was no sound, naturally.

"Blasting! We're pretty close."

"Keep your eyes open," said Powell.

Donovan nodded impatiently.

It was upon them and gone before they could seize themselves – just a bronze glint across the field of vision. They clung together in silence.

Powell whispered, "Think it sensed us?"

"Hope not. But we'd better flank them. Take the first side tunnel to the right."

"Suppose we miss them altogether?"

"Well what do you want to do? Go back?" Donovan grunted fiercely. "They're within a quarter of a mile. I was watching them through the visiplate, wasn't I? And we've got two days – "

"Oh, shut up. You're wasting your oxygen. Is this a side passage here?" The flash flicked. "It is. Let's go."

The vibration was considerably more marked and the ground below shuddered uneasily.

"This is good," said Donovan, "if it doesn't give out on us, though." He flung his light ahead anxiously.

They could touch the roof of the tunnel with a half-upstretched hand, and the bracings had been newly placed.

Donovan hesitated, "Dead end, let's go back."

"No. Hold on." Powell squeezed clumsily past. "Is that light ahead?"

"Light? I don't see any. Where would there be light down here?"

"Robot light." He was scrambling up a gentle incline on hands and knees. His voice was hoarse and anxious in Donovan's ears. "Hey, Mike, come up here."

There was light. Donovan crawled up and over Powell's outstretched legs. "An opening?"

"Yes. They must be working into this tunnel from the other side now I think."

Donovan felt the ragged edges of the opening that looked out into what the cautious flashlight showed to be a larger and obviously main stem tunnel. The hole was too small for a man to go through, almost too small for two men to look through simultaneously.

There's nothing there," said Donovan.

"Well, not now. But there must have been a second ago or we wouldn't have seen light. Watch out!"

The walls rolled about them and they felt the impact. A fine dust showered down. Powell lifted a cautious head and looked again. "All right, Mike. They're there."

The glittering robots clustered fifty feet down the main stem. Metal arms labored mightily at the rubbish heap brought down by the last blast.

Donovan urged eagerly, “Don’t waste time. It won’t be long before they get through, and the next blast may get us.”

“For Pete’s sake, don’t rush me.” Powell unlimbered the detonator, and his eyes searched anxiously across the dusky background where the only light was robot light and it was impossible to tell a projecting boulder from a shadow.

“There’s a spot in the roof, see it, almost over them. The last blast didn’t quite get it. If you can get it at the base, half the roof will cave in.”

Powell followed the dim finger, “Check! Now fasten your eye on the robots and pray they don’t move too far from that part of the tunnel. They’re my light sources. Are all seven there?”

Donovan counted, “All seven.”

“Well, then, watch them. Watch every motion!”

His detonator was lifted and remained poised while Donovan watched and cursed and blinked the sweat out of his eye.

It flashed!

There was a jar, a series of hard vibrations, and then a jarring thump that threw Powell heavily against Donovan.

Donovan yowled, “Greg, you threw me off. I didn’t see a thing.”

Powell stared about wildly, “Where are they?”

Donovan fell into a stupid silence. There was no sign of the robots. It was dark as the depths of the River Styx.

“Think we buried them?” quavered Donovan.

“Let’s get down there. Don’t ask me what I think.” Powell crawled backward at tumbling speed.

“Mike!”

Donovan paused in the act of following. “What’s wrong now?”

“Hold on!” Powell’s breathing was rough and irregular in Donovan’s ears. “Mike! Do you hear me, Mike?”

“I’m right here. What is it?”

“We’re blocked in. It wasn’t the ceiling coming down fifty feet away that knocked us over. It was our own ceiling. The shock’s tumbled it!”

“What!” Donovan scrambled up against a hard barrier. “Turn on the flash.”

Powell did so. At no point was there room for a rabbit to squeeze through.

Donovan said softly, “Well, what do you know?”

They wasted a few moments and some muscular power in an effort to move the blocking barrier. Powell varied this by wrenching at the edges of the original hole. For a moment, Powell lifted his blaster. But in those close quarters, a flash would be suicide and he knew it. He sat down.

"You know, Mike," he said, "we've really messed this up. We are no nearer finding out what's wrong with Dave. It was a good idea but it blew up in our face."

Donovan's glance was bitter with an intensity totally wasted on the darkness, "I hate to disturb you, old man, but quite apart from what we know or don't know of Dave, we're slightly trapped. If we don't get loose, fella, we're going to die. D-I-E, die. How much oxygen have we anyway? Not more than six hours."

"I've thought of that." Powell's fingers went up to his longsuffering mustache and clanged uselessly against the transparent visor. "Of course, we could get Dave to dig us out easily in that time, except that our precious emergency must have thrown him off, and his radio circuit is out."

"And isn't that nice?"

Donovan edged up to the opening and managed to get his metal incased head out. It was an extremely tight fit.

"Hey, Greg!"

"What?"

"Suppose we get Dave within twenty feet. He'll snap to normal. That will save us."

"Sure, but where is he?"

"Down the corridor – way down. For Pete's sake, stop pulling before you drag my head out of its socket. I'll give you your chance to look."

Powell maneuvered his head outside, "We did it all right. Look at those saps. That must be a ballet they're doing."

"Never mind the side remarks. Are they getting any closer?"

"Can't tell yet. They're too far away. Give me a chance. Pass me my flash, will you? I'll try to attract their attention that way."

He gave up after two minutes, "Not a chance! They must be blind. Uh-oh, they're starting toward us. What do you know?"

Donovan said, "Hey, let me see!"

There was a silent scuffle. Powell said, "All right!" and Donovan got his head out.

They were approaching. Dave was high-stepping the way in front and the six "fingers" were a weaving chorus line behind him.

Donovan marveled, "What are they doing? That's what I want to know. It looks like the Virginia reel – and Dave's a major-domo, or I never saw one."

“Oh, leave me alone with your descriptions,” grumbled Powell. “How near are they?”

“Within fifty feet and coming this way. We’ll be out in fifteen min – Uh-huh-HUH-HEY-Y!”

“What’s going on?” It took Powell several seconds to recover from his stunned astonishment at Donovan’s vocal gyrations. “Come on, give me a chance at that hole. Don’t be a hog about it.”

He fought his way upward, but Donovan kicked wildly, “They did an about-face, Greg. They’re leaving. Dave! Hey, Da-a ave!”

Powell shrieked, “What’s the use of that, you fool? Sound won’t carry.”

“Well, then,” panted Donovan, “kick the walls, slam them, get some vibration started. We’ve got to attract their attention somehow, Greg, or we’re through. ” He pounded like a madman.

Powell shook him, “Wait, Mike, wait. Listen, I’ve got an idea. Jumping Jupiter, this is a fine time to get around to the simple solutions. Mike!”

“What do you want?” Donovan pulled his head in.

“Let me in there fast before they get out of range.”

“Out of range! What are you going to do? Hey, what are you going to do with that detonator?” He grabbed Powell’s arm.

Powell shook off the grip violently. “I’m going to do a little shooting.”

“Why?”

“That’s for later. Let’s see if it works first. If it doesn’t, then – Get out of the way and let me shoot!”

The robots were flickers, small and getting smaller, in the distance. Powell lined up the sights tensely, and pulled the trigger three times. He lowered the guns and peered anxiously. One of the subsidiaries was down! There were only six gleaming figures now.

Powell called into his transmitter uncertainly. “Dave!”

A pause, then the answer sounded to both men, “Boss? Where are you? My third subsidiary has had his chest blown in. He’s out of commission.”

“Never mind your subsidiary,” said Powell. “We’re trapped in a cave-in where you were blasting. Can you see our flashlight?”

“Sure. We’ll be right there.”

Powell sat back and relaxed, “That, my fran’, is that.”

Donovan said very softly with tears in his voice, “All right, Greg. You win. I beat my forehead against the ground before your feet. Now don’t feed me any bull. Just tell me quietly what it’s all about.”

"Easy. It's just that all through we missed the obvious – as usual. We knew it was the personal initiative circuit, and that it always happened during emergencies, but we kept looking for a specific order as the cause. Why should it be an order?"

"Why not?"

"Well, look, why not a type of order. What type of order requires the most initiative? What type of order would occur almost always only in an emergency?"

"Don't ask me, dreg. Tell me!"

"I'm doing it! It's the six-way order. Under all ordinary conditions, one or more of the 'fingers' would be doing routine tasks requiring no close supervision – in the sort of offhand way our bodies handle the routine walking motions. But in an emergency, all six subsidiaries must be mobilized immediately and simultaneously. Dave must handle six robots at a time and something gives. The rest was easy. Any decrease in initiative required, such as the arrival of humans, snaps him back. So I destroyed one of the robots. When I did, he was transmitting only five-way orders. Initiative decreases – he's normal."

"How did you get all that?" demanded Donovan.

"Just logical guessing. I tried it and it worked."

The robot's voice was in their ears again, "Here I am. Can you hold out half an hour?"

"Easy!" said Powell. Then, to Donovan, he continued, "And now the job should be simple. We'll go through the circuits, and check off each part that gets an extra workout in a six-way order as against a five-way. How big a field does that leave us?"

Donovan considered, "Not much, I think. If Dave is like the preliminary model we saw back at the factory, there's a special coordinating circuit that would be the only section involved." He cheered up suddenly and amazingly, "Say, that wouldn't be bad at all. There's nothing to that."

"All right. You think it over and we'll check the blueprints when we get back. And now, till Daves reaches us, I'm relaxing."

"Hey, wait! Just tell me one thing. What were those queer shifting marches, those funny dance steps, that the robots went through every time they went screwy?"

"That? I don't know. But I've got a notion. Remember, those subsidiaries were Dave's 'fingers.' We were always saying that, you know. Well, it's my idea that in all these interludes, whenever Dave became a psychiatric case, he went off into a moronic maze, spending his time twiddling his fingers."

After You Have Read the Story …

"Catch That Rabbit" reflects the challenge of debugging a robot that has sophisticated hardware and software. The problem with the DV-5 robot in the story is determined to be hardware associated with the switching mechanisms needed to control multiple robots. It is interesting that in the story, the positronic brain (software) is presumed to work correctly and thus it must be the hardware malfunctioning. This is generally the opposite of real life, as the software is immediately suspected as the cause of any problems. Hardware is easier to test than software.

The story is a springboard for discussing testing and autonomy in more detail. It ignores that user trust in a robot actually working as expected is a major factor in whether a robot is adopted by users; the robot passes the V&V tests but the users quickly stop using it because they don't trust it. There are two aspects to trusting a robot: what level of initiative is delegated to the robot and whether the rationale of the robot for its actions is visible. The need to produce trustworthy systems is changing the landscape of robotics testing from verification and validation to verification, validation, and visibility. The odd part of the story is the division of responsibilities between DV-5 and the rest of the robots, but this fortunately leads to an exploration of how autonomy is related to the initiative that an agent is allowed to exercise. For more in-depth reading, levels of initiative are covered in chapter 4, trust and visibility in chapter 18, and testing in chapter 19 of *Introduction to AI Robotics*;[2] see chapter 17 for readings in multirobot systems or the classic book *Robot Teams: From Diversity to Polymorphism*.[3]

Testing Nondeterministic Systems

In theory, nondeterminism might be deterministically modeled with supercomputers but would require extensive effort. If a person is asked to take the exit to the left, it generally doesn't matter if they take fifteen steps or twelve steps to reach the exit, just that they follow the directions. Furthermore, the number of steps may vary if they are feeling sick or are eager to comply, so predicting the exact number of steps a person would take would require a very extensive model. Worse yet, walking to an exit has many possible paths—one person may favor the right side of a corridor and another the center—but at least the number of steps is discrete and has a fixed sequence, so it is conceivable that a person's actions could be modeled.

Now imagine trying to model how a person will open a door, when each person may apply a different combination of grasping configurations and grip strength in a unique sequence based on what they learned as a child. The grasping position and grip strength are continuous variables, covering a range of values, rather than a discrete, or binary, count of number of steps walked. This makes modeling even harder.

In human testing, there is no need to apply super-computing to predict the number of steps they will take for each situation—instead testing concentrates on measuring how likely a person is to accomplish a task. The predictions of human behavior become statistical, for example that 99.9 percent of the population will open and exit an airlock door within ninety seconds. This avoids modeling nondeterministic conditions. As a result, statistical measures of performance are key for artificial intelligence robotics: does the robot do the right thing overall even though there may be variations in how it does it? This is quite different than deterministically treating the robot as a machine like an automobile, elevator, or industrial robot arm. At best, deterministic testing provides verification of individual specifications, not validation that the overall performance of the robot is acceptable.

The story is realistic in how Donovan and Powell rely on *regression testing,* in which the tester subjects the system to a set of known conditions and confirms that it performs the same way (ideally the original way was the right way, but regression testing only guarantees consistency). This type of testing is important in software engineering because it helps establish that a new feature doesn't interfere with existing features. Still, the cases for regression testing are limited by the imagination of the designer and thus may not represent what later turns out to be an important confluence of variables. Regression testing is not sufficient for nondeterministic systems because the test cases may not capture very subtle or rare conditions.

Trust and Visibility

In real life, testing is also important for building user trust in a system. Trust is indirectly referred to within the story as a concern over the robotics company's reputation and future sales. But it is a big issue because users will discard or avoid using robots that make them uncomfortable. For example, US soldiers and officers simply avoided using robots that they didn't trust during the second Gulf War.[4]

Although testing is necessary to build a user's personal trust in a system, it is not sufficient. Another key element in building user trust is *visibility*, which also helps with debugging. Visibility is a concept from user interfaces that acknowledges that people don't trust "black boxes"; they want to see or have some indication of what a system is doing. The spinning rainbow wheel or hour glass that comes up on computers to show that a process is still running is an example of visibility; because the user has an indication that the computer is doing something it is supposed to be doing, the user trusts that the computer has not failed and is willing to wait until that process has completed.

The lack of visibility in how each robot is working clearly hinders debugging in the story, and it does not reflect what are now known as best practices in engineering and AI. First, the robots do not have any "black boxes" such as those used on aircraft and automobiles that record the state of key inputs and outputs for accident investigation. Second, the need for an artificial intelligence to be able to display the cause for its actions or decisions has been known since 1977 when the TEIRESIAS was created to explain to doctors why the MYCIN medical diagnosis expert system had come to the conclusion it had.[5] An advantage of a medical expert system is that it can think of causes that doctors simply don't think of or have prematurely dismissed. But doctors were reluctant to order expensive or painful tests just because a computer said so. The doctors were also aware that an expert system might not have all the information in its knowledge base that they had as human practitioners. TEIRESIAS allowed doctors to see what the computer was thinking and then, if they found a problem with the program's reasoning, assisted the doctors in making any updates.

Autonomy and Initiative

In "Catch That Rabbit," Asimov frames his robot's problem as one of personal initiative since the level of initiative needed by the DV-5 robot is exceeded during an emergency with six subordinate robots. In AI robotics, one way of designing autonomy is to consider the level of initiative needed to appropriately accomplish a task. Alan Colman and Jun Han were among the first roboticists to explore initiative as a way of thinking about autonomy between team members.[6] In applications such as RoboCup soccer, which was their motivating domain, the challenge is how much freedom the individual robots on the team have to deviate from play. They came up with five types of initiative. One type is *no autonomy*, where all the robots strictly adhere to their role in accomplishing the goal (or soccer play), or, in this story, the directives given to them by DV-5. In *process autonomy* and *systems-state autonomy*, the individual robots are given a goal and various freedoms to choose how to fulfill their role in achieving it. Returning to the deliberative functions discussed with "Long Shot," DV-5 generates and presumably monitors a plan but his subordinate robots would be allowed to select and implement their own behaviors for accomplishing their part of the play. With *intentional autonomy*, the individual robots have flexibility in changing their part of the play in order to better meet the intent of the team goal. If one of the robots is getting behind in its task, another robot might decide on its own to go help out. The fifth type of initiative is *constraint autonomy*, where the robot is allowed to discard constraints. In AI, this is usually meant in terms of "relaxing" a constraint on planning—for example, "I can't do all three tasks on my to-do list, what if I just try to do two of them?" Of course, constraint autonomy raises the question of robots reducing constraints and taking over the world. Fortunately, constraint autonomy is still subject to the robot's *bounded rationality*—so unless it was programmed to take over the world and had access to all the necessary resources needed to take over the world, it wouldn't.

In the story, DV-5 is a centralized controller and the six robots have *no autonomy*, which overloads the switching system. In real life, the centralization would not have stressed the hardware per se but rather the software. Increasing the number of robots from five to six and then adding an emergency is an example of increasing the amount of computation needed to plan and coordinate the robots. This amount of computation is referred to as *computational complexity*. Algorithms are formally rated in terms of how much computation they require for different inputs; for example, an algorithm may require one hundred operations to coordinate two robots, but one thousand operations to control three, ten thousand to control four, and so on. In computer science parlance, the cost, also known as order complexity *O*, of the algorithm is n^2, where n is the number of key operations, abbreviated as $O(n^2)$. So it is feasible that an algorithm that runs well on a particular processor for five robots may fail with six because the increase in computation that has to be completed in real time might finally exceed the threshold of the processor. In this story, the failure is more a design mismatch between the algorithm and the processor, not a mechanical failure—though in real life changing processors might have fixed the problem. Computational complexity problems typically lead to degraded operations. In a computationally bounded situation, it is more likely that the robots would still perform their tasks but in a poorly coordinated and jerky fashion because the right commands weren't being computed fast enough. They wouldn't be coming up with alternative tasks, as DV-5 does when the robots start marching in formation.

The computational or switching complexity that overwhelms DV-5 might have been mitigated by allowing the robots to have at least process or systems-state autonomy so that they do not have to depend on DV-5 for explicit commands. Each robot would understand the goal (possibly captured by the *script* procedural knowledge representation discussed with "Long Shot") and their role in achieving it and simply execute their local strategies. DV-5 would be more of a quarterback or coach announcing plays and even shouting changes as needed.

Reality Score: A–

"Catch That Rabbit" gets major points for foreseeing the challenges of testing and evaluation and how invisible states interfere with debugging. Even now, testing, evaluation, and debugging are too often afterthoughts in design, despite known software engineering and human-computer interaction principles. The story misses a perfect rating due to the unrealistic, but amusing, centralized control and coordination regime.

As another example of life imitating art, a "Handbook" of robotics does exist, though not as the ultimate debugging manual envisioned by Asimov. The *Springer Handbook of Robotics* was first published in 2008 and a second edition in 2016.[7] It was the creation of Bruno Siciliano, Oussama Khatib, and over sixty other leading researchers and supplies a comprehensive guide to the state of knowledge for all aspects of robotics. I was the lead author on the chapter on search and rescue robots. The *Handbook* won the AAP PROSE Award for Excellence in Physical Sciences & Mathematics as well as the Award for Engineering & Technology, which should reassure the reader that the volume is readable.

5 Human-Robot Interaction: "Supertoys Last All Summer Long"

"Supertoys Last All Summer Long," written in 1969 by Brian Aldiss, weaves several threads together in a vision of a possible near future: overpopulation, the advent of artificially intelligent robots, and individual isolation and despair.[1] It was the source material for the 2001 movie *A.I. Artificial Intelligence*, which was first put into production by Stanley Kubrick and then taken over by Steven Spielberg when Kubrick died, which could explain the significant differences between the story and the movie.

The plot centers on Monica, her three-year-old son David, David's intelligent teddy bear Teddy, and Monica's husband, who is the managing director of Synthank. Synthank has moved from creating tape worms that keep people slim (presaging Mira Grant's Parasitology science fiction series) to servant robots, promising to have the equivalent of an Einstein in every home. The technological MacGuffin in this story is that robots must be socially aware to fit in with the norms of everyday home life. Socially aware robots, and how robots fit in with all types of people and in all types of circumstances, are addressed by the subdiscipline of robotics called *human-robot interaction*.

As You Read the Story ...

Human-robot interaction is often confused with creating naturalistic user interfaces to better direct robots. Users often ask for dialog-based interfaces—for example, "I just want to tell the robot what to do" or "I want the robot to explain to me why it is doing that." They seem surprised, and even annoyed, that roboticists don't already provide these types of naturalistic interfaces, especially with the advent of programs such as Siri.

This focus on interfaces neglects other forms of communication, such as how people look and act, and the myriad types of relationships we can have with each other.

As you read the story, mark the ways in which a young boy is having difficulties with language and expressing himself to others, even with Teddy's help. One method of communication is verbal dialog, or two-way communication, though people and animals also express their intents and emotional states through gestures, eye contact, and body posture. People also write, which is one-way communication, and create visual arts, such as David's drawings, to communicate. Keep track of what Teddy says and does and compare it to what David says and does.

"Supertoys Last All Summer Long" by Brian Aldiss, 1969

In Mrs. Swinton's garden, it was always summer. The lovely almond trees stood about it in perpetual leaf. Monica Swinton plucked a saffron-colored rose and showed it to David.

"Isn't it lovely?" she said.

David looked up at her and grinned without replying. Seizing the flower, he ran with it across the lawn and disappeared behind the kennel where the mowervator crouched, ready to cut or sweep or roll when the moment dictated. She stood alone on her impeccable plastic gravel path.

She had tried to love him.

When she made up her mind to follow the boy, she found him in the courtyard floating the rose in his paddling pool. He stood in the pool engrossed, still wearing his sandals.

"David, darling, do you have to be so awful? Come in at once and change your shoes and socks."

He went with her without protest into the house, his dark head bobbing at the level of her waist. At the age of three, he showed no fear of the ultrasonic dryer in the kitchen. But before his mother could reach for a pair of slippers, he wriggled away and was gone into the silence of the house.

He would probably be looking for Teddy.

Monica Swinton, twenty-nine, of graceful shape and lambent eye, went and sat in her living room, arranging her limbs with taste. She began by sitting and thinking; soon she was just sitting. Time waited on her shoulder with the maniac slowth it reserves for children, the insane, and wives whose husbands are away improving the world. Almost by reflex, she reached out and changed the wavelength of her windows. The garden faded; in its place, the city center rose by her left hand, full of crowding people, blowboats, and buildings (but she kept the sound down). She remained alone. An overcrowded world is the ideal place in which to be lonely.

The directors of Synthank were eating an enormous luncheon to celebrate the launching of their new product. Some of them wore the plastic face-masks popular at the time. All were elegantly slender, despite the rich food and drink they were putting away. Their wives were elegantly slender, despite the food and drink they too were putting away. An earlier and less sophisticated generation would have regarded them as beautiful people, apart from their eyes.

Henry Swinton, Managing Director of Synthank, was about to make a speech.

"I'm sorry your wife couldn't be with us to hear you," his neighbor said.

"Monica prefers to stay at home thinking beautiful thoughts," said Swinton, maintaining a smile.

"One would expect such a beautiful woman to have beautiful thoughts," said the neighbor.

Take your mind off my wife, you bastard, thought Swinton, still smiling.

He rose to make his speech amid applause.

After a couple of jokes, he said, "Today marks a real breakthrough for the company. It is now almost ten years since we put our first synthetic life-forms on the world market. You all know what a success they have been, particularly the miniature dinosaurs. But none of them had intelligence.

"It seems like a paradox that in this day and age we can create life but not intelligence. Our first selling line, the Crosswell Tape, sells best of all, and is the most stupid of all." Everyone laughed.

"Though three-quarters of the overcrowded world are starving, we are lucky here to have more than enough, thanks to population control. Obesity's our problem, not malnutrition. I guess there's nobody round this table who doesn't have a Crosswell working for him in the small intestine, a perfectly safe parasite tape-worm that enables its host to eat up to fifty percent more food and still keep his or her figure. Right?" General nods of agreement.

"Our miniature dinosaurs are almost equally stupid. Today, we launch an intelligent synthetic life-form – a full-size serving-man.

"Not only does he have intelligence, he has a controlled amount of intelligence. We believe people would be afraid of a being with a human brain. Our serving-man has a small computer in his cranium.

"There have been mechanicals on the market with mini-computers for brains – plastic things without life, super-toys – but we have at last found a way to link computer circuitry with synthetic flesh."

David sat by the long window of his nursery, wrestling with paper and pencil. Finally, he stopped writing and began to roll the pencil up and down the slope of the desk-lid.

"Teddy!" he said.

Teddy lay on the bed against the wall, under a book with moving pictures and a giant plastic soldier. The speech-pattern of his master's voice activated him and he sat up.

"Teddy, I can't think what to say!"

Climbing off the bed, the bear walked stiffly over to cling to the boy's leg. David lifted him and set him on the desk.

"What have you said so far?"

"I've said – " He picked up his letter and stared hard at it. "I've said, 'Dear Mummy, I hope you're well just now. I love you. ...'"

There was a long silence, until the bear said, "That sounds fine. Go downstairs and give it to her."

Another long silence.

"It isn't quite right. She won't understand."

Inside the bear, a small computer worked through its program of possibilities. "Why not do it again in crayon?"

When David did not answer, the bear repeated his suggestion. "Why not do it again in crayon?"

David was staring out of the window. "Teddy, you know what I was thinking? How do you tell what are real things from what aren't real things?"

The bear shuffled its alternatives. "Real things are good."

"I wonder if time is good. I don't think Mummy likes time very much. The other day, lots of days ago, she said that time went by her. Is time real, Teddy?"

"Clocks tell the time. Clocks are real. Mummy has clocks so she must like them. She has a clock on her wrist next to her dial."

David started to draw a jumbo jet on the back of his letter. "You and I are real, Teddy, aren't we?"

The bear's eyes regarded the boy unflinchingly. "You and I are real, David." It specialized in comfort.

Monica walked slowly about the house. It was almost time for the afternoon post to come over the wire. She punched the Post Office number on the dial on her wrist but nothing came through. A few minutes more.

She could take up her painting. Or she could dial her friends. Or she could wait till Henry came home. Or she could go up and play with David. ...

She walked out into the hall and to the bottom of the stairs.

"David!"

No answer. She called again and a third time.

"Teddy!" she called, in sharper tones.

"Yes, Mummy!" After a moment's pause, Teddy's head of golden fur appeared at the top of the stairs.

"Is David in his room, Teddy?"

"David went into the garden, Mummy."

"Come down here, Teddy!"

She stood impassively, watching the little furry figure as it climbed down from step to step on its stubby limbs. When it reached the bottom, she picked it up and carried it into the living room. It lay unmoving in her arms, staring up at her. She could feel just the slightest vibration from its motor.

"Stand there, Teddy. I want to talk to you." She set him down on a tabletop, and he stood as she requested, arms set forward and open in the eternal gesture of embrace.

"Teddy, did David tell you to tell me he had gone into the garden?"

The circuits of the bear's brain were too simple for artifice. "Yes, Mummy."

"So you lied to me."

"Yes, Mummy."

"Stop calling me Mummy! Why is David avoiding me? He's not afraid of me, is he?"

"No. He loves you."

"Why can't we communicate?"

"David's upstairs."

The answer stopped her dead. Why waste time talking to this machine? Why not simply go upstairs and scoop David into her arms and talk to him, as a loving mother should to a loving son? She heard the sheer weight of silence in the house, with a different quality of silence pouring out of every room. On the upper landing, something was moving very silently – David, trying to hide away from her. ...

He was nearing the end of his speech now. The guests were attentive; so was the Press, lining two walls of the banqueting chamber, recording Henry's words and occasionally photographing him.

"Our serving-man will be, in many senses, a product of the computer. Without computers, we could never have worked through the sophisticated biochemics that go into synthetic flesh. The serving-man will also be an extension of the computer – for he will contain a computer in his own head, a microminiaturized computer capable of dealing with almost any situation he may encounter in the home. With reservations, of course." Laughter at this; many of those present knew the heated debate that had engulfed the Synthank boardroom before the decision had finally been taken to leave the serving-man neuter under his flawless uniform.

"Amid all the triumphs of our civilization – yes, and amid the crushing problems of overpopulation too – it is sad to reflect how many millions of people suffer from increasing loneliness and isolation. Our serving-man will be a boon to them; he will always answer, and the most vapid conversation cannot bore him.

"For the future, we plan more models, male and female – some of them without the limitations of this first one, I promise you! – of more advanced design, true bio-electronic beings.

"Not only will they possess their own computer, capable of individual programming; they will be linked to the World Data Network. Thus everyone will be able to enjoy the equivalent of an Einstein in their own homes. Personal isolation will then be banished forever!"

He sat down to enthusiastic applause. Even the synthetic serving-man, sitting at the table dressed in an unostentatious suit, applauded with gusto.

Dragging his satchel, David crept round the side of the house. He climbed on to the ornamental seat under the living-room window and peeped cautiously in.

His mother stood in the middle of the room. Her face was blank; its lack of expression scared him. He watched fascinated. He did not move; she did not move. Time might have stopped, as it had stopped in the garden.

At last she turned and left the room. After waiting a moment, David tapped on the window. Teddy looked round, saw him, tumbled off the table, and came over to the window. Fumbling with his paws, he eventually got it open.

They looked at each other.

"I'm no good, Teddy. Let's run away!"

"You're a very good boy. Your Mummy loves you."

Slowly, he shook his head. "If she loved me, then why can't I talk to her?"

"You're being silly, David. Mummy's lonely. That's why she had you."

"She's got Daddy. I've got nobody 'cept you, and I'm lonely."

Teddy gave him a friendly cuff over the head. "If you feel so bad, you'd better go to the psychiatrist again."

"I hate that old psychiatrist – he makes me feel I'm not real." He started to run across the lawn. The bear toppled out of the window and followed as fast as its stubby legs would allow.

Monica Swinton was up in the nursery. She called to her son once and then stood there, undecided. All was silent.

Crayons lay on his desk. Obeying a sudden impulse, she went over to the desk and opened it. Dozens of pieces of paper lay inside. Many of them were written in crayon in David's clumsy writing, with each letter picked out in a color different from the letter preceding it. None of the messages was finished.

"My dear Mummy, How are you really, do you love me as much – "

"Dear Mummy, I love you and Daddy and the sun is shining – "

"Dear dear Mummy, Teddy's helping me write to you. I love you and Teddy – "

"Darling Mummy, I'm your one and only son and I love you so much that some times – "

"Dear Mummy, you're really my Mummy and I hate Teddy – "

"Darling Mummy, guess how much I love – "

"Dear Mummy, I'm your little boy not Teddy and I love you but Teddy – "

"Dear Mummy, this is a letter to you just to say how much how ever so much – "

Monica dropped the pieces of paper and burst out crying. In their gay inaccurate colors, the letters fanned out and settled on the floor.

Henry Swinton caught the express home in high spirits, and occasionally said a word to the synthetic serving-man he was taking home with him. The serving-man answered politely and punctually, although his answers were not always entirely relevant by human standards.

The Swintons lived in one of the ritziest city-blocks, half a kilometer above the ground. Embedded in other apartments, their apartment had no windows to the outside; nobody wanted to see the overcrowded external world. Henry unlocked the door with his retina pattern-scanner and walked in, followed by the serving-man.

At once, Henry was surrounded by the friendly illusion of gardens set in eternal summer. It was amazing what Whologram could do to create huge mirages in small spaces. Behind its roses and wisteria stood their house; the deception was complete: a Georgian mansion appeared to welcome him.

"How do you like it?" he asked the serving-man.

"Roses occasionally suffer from black spot."

"These roses are guaranteed free from any imperfections."

"It is always advisable to purchase goods with guarantees, even if they cost slightly more."

"Thanks for the information," Henry said dryly. Synthetic life-forms were less than ten years old, the old android mechanicals less than sixteen; the faults of their systems were still being ironed out, year by year.

He opened the door and called to Monica.

She came out of the sitting-room immediately and flung her arms round him, kissing him ardently on cheek and lips. Henry was amazed.

Pulling back to look at her face, he saw how she seemed to generate light and beauty. It was months since he had seen her so excited. Instinctively, he clasped her tighter.

"Darling, what's happened?"

"Henry, Henry – oh, my darling, I was in despair ... but I've just dialed the afternoon post and – you'll never believe it! Oh, it's wonderful!"

"For heaven's sake, woman, what's wonderful?"

He caught a glimpse of the heading on the photostat in her hand, still moist from the wall-receiver: Ministry of Population. He felt the color drain from his face in sudden shock and hope.

"Monica ... oh ... Don't tell me our number's come up!"

"Yes, my darling, yes, we've won this week's parenthood lottery! We can go ahead and conceive a child at once!"

He let out a yell of joy. They danced round the room. Pressure of population was such that reproduction had to be strict, controlled. Childbirth required government permission. For this moment, they had waited four years. Incoherently they cried their delight.

They paused at last, gasping, and stood in the middle of the room to laugh at each other's happiness. When she had come down from the nursery, Monica had de-opaqued the windows, so that they now revealed the vista of garden beyond. Artificial sunlight was growing long and golden across the lawn – and David and Teddy were staring through the window at them.

Seeing their faces, Henry and his wife grew serious.

"What do we do about them?" Henry asked.

"Teddy's no trouble. He works well."

"Is David malfunctioning?"

"His verbal communication-center is still giving trouble. I think he'll have to go back to the factory again."

"Okay. We'll see how he does before the baby's born. Which reminds me – I have a surprise for you: help just when help is needed! Come into the hall and see what I've got."

As the two adults disappeared from the room, boy and bear sat down beneath the standard roses.

"Teddy – I suppose Mummy and Daddy are real, aren't they?"

Teddy said, "You ask such silly questions, David. Nobody knows what 'real' really means. Let's go indoors."

"First I'm going to have another rose!" Plucking a bright pink flower, he carried it with him into the house. It could lie on the pillow as he went to sleep. Its beauty and softness reminded him of Mummy.

After You Have Read the Story ...

"Stranger in Paradise" always brings tears to my eyes because of the autism theme, but "Supertoys Last All Summer Long" triggers a flood. Spielberg's movie is quite different from the story, though Teddy appears relatively unchanged. In the movie, David eventually finds acceptance from Monica but the story suggests a much grimmer ending. In the story, David will likely remain switched-on but will exist as a discarded supertoy, enduring as long as the plastic gravel in Mrs. Swinton's always summer garden and garnering as little consideration.

The story is an emotional introduction to key concepts in human-robot interaction, discussed below and in chapter 18 (on human-robot interaction) of *Introduction to AI Robotics*.[2] The AI community has historically used the Turing test, in which a computer is hidden and must express its intelligence and engage with users through natural language, as the means for determining intelligence in computers. More recently, roboticists have begun to use the uncanny valley, where the robot is visible, for determining acceptability. The uncanny valley, which will be described in more detail in the next section, is essentially a test of whether the robot makes the human observer uncomfortable because it is just lifelike enough to be eerie or uncanny. Both Teddy and David would pass the Turing test easily, and David does not fall in the uncanny valley.

Cleverly, Aldiss exploits linguistic ambiguity to set up the gut-wrenching emotional twist at the end of the story. Given the clear disconnect between Monica and David, the story is a great illustration of ideas in AI such as common ground and the belief-desires-intention framework. Because Monica has made assumptions about David's capabilities as a robot, Monica doesn't view David's letters and drawings as real communication and ignores David's beliefs, desires, and intentions. Meanwhile, David struggles to infer Monica's beliefs, desires, and intentions within his limited model that she is his mother and is acting in accordance to that role. Just like a "real" three-year-old, David can't figure out why his mother isn't acting as a mother and blames himself.

Turing Test and the Uncanny Valley

If it is hard for people to communicate with each other, imagine the challenges for people and robots. Many researchers believe natural language is the epitome of artificial intelligence because of the breadth of knowledge that must be represented and the ability to infer meaning from verbal and nonverbal cues in order to follow a conversation. This is perhaps one reason why the Turing test remains the de facto standard for general intelligence.[3] The Turing test states that if a real person has a typed dialog with a computer (typed so the identity of the conversant, whether a person or machine, is hidden) and can't tell if it is a human or a computer, the computer is intelligent. If the computer is so knowledgeable about the world and socially aware that it is able to respond quickly and appropriately in a conversation with a human, then it would have to be considered intelligent. The Turing test has been hijacked by various "chat-bot" competitions in which different AI systems appear to be human more because of their cantankerous attitude and refusal to docilely follow conversational threads than their conversational ability.

The Turing test was proposed in 1950 by Alan Turing, who wanted to hide the identity of the machine; this is one of the themes in *The Imitation Game,* the 2014 movie about Turing and his homosexuality. But robots are visible, which leads to a different, additional perspective on intelligence called the uncanny valley. The uncanny valley concept emerged in 1979, one year after "Supertoys Last All Summer Long" was published, when Masahiro Mori described a graph of a robot's physical likeness versus a human's comfort level with the robot. This graph is reproduced and updated in figure 5.1. It had a dip in the middle dubbed the uncanny valley, for which Mori hypothesized that robots that appeared lifelike but did not have commensurate lifelike movements or expressions would be considered creepy.[4] On one side of the valley, a robot is acceptable if it looks clunky and reacts slowly because it isn't very smart. The robot doesn't look very much like a human, so the fact that it doesn't quite act like one is fine. On the other side of the valley, a robot that looks and moves exactly like a human is acceptable. The problem occurs in the middle ground, or valley, when a robot is either physically misproportioned such as Diego-san, generally considered the world's creepiest robot, or looks very human but doesn't quite move correctly or have the right facial expressions.[5] The more human-like the robot, the less we tolerate it not quite moving or acting like a human. Being perfectly still is acceptable in a bomb squad robot that looks like a tank but disquieting in an android robot such as those in the TV series *Westworld* or *Humans*.

In "Supertoys Last All Summer Long," David apparently looks and acts like a perfect, life-like child, with nothing beyond the label of being a

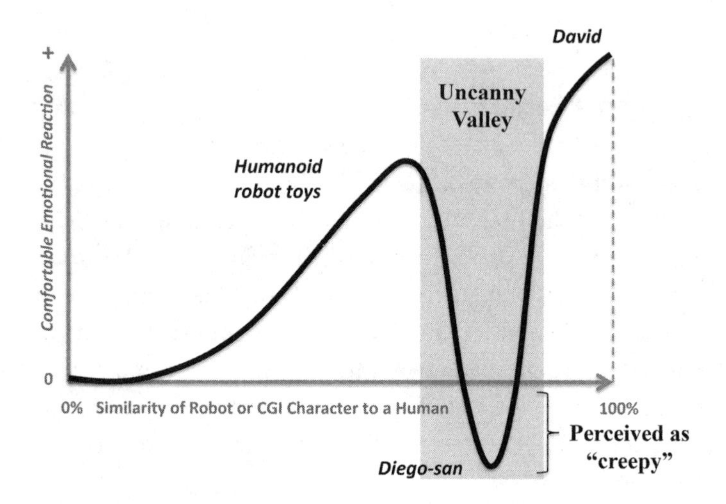

Figure 5.1
The uncanny valley (after Mori, "The Uncanny Valley").

"robot" to cause unease. He passes the Turing test and is on an acceptable side of the uncanny valley. In theory, it would be hard for a person not to treat David like a boy.

Natural Language Understanding

Understanding natural language is challenging for many reasons. One is ambiguity, which occurs in even simple commands such as telling a robot to "go over there." "Over there" means where exactly? Pointing, also called *deitic gesturing,* can help resolve this ambiguity but it expands the computational effort required to include computer vision and adds the problem of associating a gesture with a verbal instruction. It introduces spatial uncertainty as well as the need to learn individual styles in gesturing. Another problem in natural language understanding is context. Two people in a pleasant dialog might make ironic compliments such as "you suck," but the same phrase might be an insult in a different context.

There are three general approaches to understanding natural language: to focus on the *words,* the *sentence structure,* or the *underlying semantics* or meaning. Voice activation programs on phones are an impressive example of natural language understanding that focuses primarily on words and sentence structure. Those programs take advantage of the finite number of things you can do with your smart phone, which can be expressed with a restricted lexicon of words, and uses those restrictions to infer intent and help with ambiguity and context. Also, people naturally speak in simple, short sentences to machines, making it easier to extract what they want.

Belief-Desire-Intention Model of AI

The Belief-Desire-Intention (BDI) model of AI grew out of the challenges in natural language. Understanding the underlying semantics or meaning of a dialog is much harder than the other two approaches to understanding natural language because conversation has meanings and information that are implicit, that is, not explicitly stated. There are two main methods of making the implicit more explicit. One is to create and maintain a model of *common ground,* which is in essence an understanding that you and I are taking about the same thing or situation. Once I have established the common ground for our conversation, I can use my internal explicit representation of that thing or situation to follow along with what you are saying. Common ground is helpful for situations in which context is important: "I think I left it in the kitchen, look in there" or "I'm hungry, can you get me something?" both mean you must go to the kitchen even though neither is stated as a command to go to the kitchen.

Another way to infer the underlying meaning of a conversation is to create a model of the *beliefs, desires, and intentions* of the other conversant.[6] A BDI approach would entail trying to internally answer questions such as "why are they talking to me?," "what do they want overall?," and "what are their expectations for this task?" in order to understand what is really being said or asked for. A BDI model is helpful for understanding that "go over there" means "go over there and don't hit anything on your way." Common ground models are generally used to follow conversation, whereas a BDI model is used to try to better understand another agent's overall intent, actions, and purposes in doing what they are doing.

Common ground and BDI intuitively appear to be essential in natural language understanding. Consider the occurrence of these two phrases in a sequence:

–Set a reminder to meet with Allan at 4 p.m.
–Send him a text message saying I'll see him this afternoon

The use of "meet" and "see him" set the common ground that someone is talking about a meeting with Allan. In addition, we have a belief that both parties already know where the meeting will be. The overall intent is unknown, but if it were known that Allan was a relative, we would know it may be social, if Allan were a coworker, it may be business, and so on.

As seen with smartphone voice interfaces, however, common ground and BDI are not necessary for executing the correct commands. Probabilistic methods focusing on words and sentence structure can associate "him" with "Allan" without understanding the underlying meaning at all and send the text to the Allan in the contact list who most frequently gets sent texts. The difference between commercially available assistants such as Siri, Alexa, and Cortana and the general artificial intelligence in a movie like *Her* is quite large.

Regardless of what underlying model of intelligence David is programmed with, he is clearly not considered intelligent by Monica. This brings to mind a brilliant lecture given at the 2001 workshop that created the field of human-robot interaction by Clifford Nass on the power of labels: "Ecce Homo: Why It's Great to Be Labeled a 'Person.'"[7] He discussed artificially intelligent robots in terms of the Dred Scott decision. Recall that Dred Scott was a slave who sued for his freedom with compelling arguments as to his personhood, but the Supreme Court ruled that since Scott was black, he was not a citizen, and since he was not a citizen, he could not sue, so the quality of his arguments didn't matter. Essentially slaves were not labeled as "people" and thus they had no rights—but more subtly, people labeled "people" had no obligation or motivation to treat slaves as people because the slaves were not labeled "people." Nass's conclusion was that we will not treat robots as intelligent beings as long as they are labeled "robots," regardless of whether they are intelligent or not. That seems to be Aldiss's conclusion as well. Having been labeled a "robot," David will always be a supertoy and never worthy of serious consideration, much less love.

As Score: A

"Supertoys Last All Summer Long" gets near perfect marks for identifying the challenge of communication, illustrating why modeling the beliefs, desires, and intentions of another agent is essential to understanding, and capturing concerns in social interactions between humans and robots. Whereas Ilse from "Long Shot" is an example of how to build an intelligent robot, David is an example of why it might be wrong to do so.

6 Ethics and Weaponization of Robots: "Second Variety"

Philip K. Dick wrote "Second Variety," the quintessential story of killer robots, in 1953.[1] Whereas Asimov's stories aim for entertainment and often leave the reader laughing, Dick's work is dark, paranoid, and often leaves the reader feeling like they have received a kick to the gut. In Dick's stories, drug dealers inform on themselves (*A Scanner Darkly*), people do not know they are literally robots (*Do Androids Dream of Electric Sheep?*), and a civilization lives in the "bad" version of the multiverse with only glimpses of how much better life could be (the 1963 Hugo Award–winning *The Man in the High Castle*).

What better writer than Dick to help us explore the ideas of the weaponization of artificially intelligent robots and self-replicating machines? The setting for "Second Variety" is a postapocalyptic earth where a war between the USA and Russia continues to rage. Small behavior-based robots called "claws" roam in swarms killing every warm creature without a radiation tag. They are built in fully roboticized factories underground and designed to protect the isolated forces that are loosely directed from generals on the moon. The technological MacGuffin is that the designers on the moon have programmed the killer robots to *learn* and to self-modify in order to keep up with the changing countertactics of the Russians. Besides machine learning, the story indirectly raises two other topics: the ethics of weaponized autonomy and bounded rationality.

As You Read the Story …

The weaponization of robots is a controversial topic. In May 2014, the United Nations held a Convention on Certain Conventional Weapons (CCW) Meeting of Experts on Lethal Autonomous Weapons Systems to discuss killer robots.[2] The focus was on whether it was acceptable to delegate decisions about lethal force to a robot. The possibility of a robot committing a war crime has been hotly debated. Ronald Arkin has argued that robots can be programmed to follow the Geneva Convention and thus are more trustworthy than humans.[3] But this assumes that robots can be deployed without unintended consequences or that any unintended consequences wouldn't impact protected groups such as civilians. In war crimes, the issue would be who made the decision to delegate lethal force to a robot and what bounds were placed on the robot.

As you read, consider the following three aspects of weaponization. First, what are the advantages of weaponized robot "claws" and are these the same arguments used for the real-life weaponization of unmanned systems? Some pro-weaponization arguments are that armed robots reduce manpower, that they are more vigilant and persistent and thus more effective at deterring enemy attacks and responding to them, that robots allow soldiers to fight remotely without the costs (or risks) of being located in the country under attack, and that if we don't do it, "they" will. Second, consider whether the designers can be absolutely sure that the claws will not spread from military bases into areas where any remaining civilians live and then kill them because they do not have radiation tags. Finally, who would be liable for war crimes in the story? Would it be the military decision makers who deployed the robots, the company or lab that built the robots, or both? Regardless of war crimes, the larger question is accountability—are the military and companies accountable for deploying a technology that could lead to a war crime, and why didn't the designers or the government stop them?

"Second Variety" by Philip K. Dick, 1953

The claws were bad enough in the first place – nasty, crawling little death-robots. But when they began to imitate their creators, it was time for the human race to make peace – if it could!

The Russian soldier made his way nervously up the ragged side of the hill, holding his gun ready. He glanced around him, licking his dry lips, his face set. From time to time he reached up a gloved hand and wiped perspiration from his neck, pushing down his coat collar.

Eric turned to Corporal Leone. "Want him? Or can I have him?" He adjusted the view sight so the Russian's features squarely filled the glass, the lines cutting across his hard, somber features.

Leone considered. The Russian was close, moving rapidly, almost running. "Don't fire. Wait." Leone tensed. "I don't think we're needed."

The Russian increased his pace, kicking ash and piles of debris out of his way. He reached the top of the hill and stopped, panting, staring around him. The sky was overcast, drifting clouds of gray particles. Bare trunks of trees jutted up occasionally; the ground was level and bare, rubble-strewn, with the ruins of buildings standing out here and there like yellowing skulls.

The Russian was uneasy. He knew something was wrong. He started down the hill. Now he was only a few paces from the bunker. Eric was getting fidgety. He played with his pistol, glancing at Leone.

"Don't worry," Leone said. "He won't get here. They'll take care of him."

"Are you sure? He's got damn far."

"They hang around close to the bunker. He's getting into the bad part. Get set!"

The Russian began to hurry, sliding down the hill, his boots sinking into the heaps of gray ash, trying to keep his gun up. He stopped for a moment, lifting his fieldglasses to his face.

"He's looking right at us," Eric said.

The Russian came on. They could see his eyes, like two blue stones. His mouth was open a little. He needed a shave; his chin was stubbled. On one bony cheek was a square of tape, showing blue at the edge. A fungoid spot. His coat was muddy and torn. One glove was missing. As he ran his belt counter bounced up and down against him.

Leone touched Eric's arm. "Here one comes."

Across the ground something small and metallic came, flashing in the dull sunlight of mid-day. A metal sphere. It raced up the hill after the Russian, its treads flying. It was small, one of the baby ones. Its claws were out, two razor projections spinning in a blur of white steel. The Russian heard it. He turned instantly, firing. The sphere dissolved into particles. But already a second had emerged and was following the first. The Russian fired again.

A third sphere leaped up the Russian's leg, clicking and whirring. It jumped to the shoulder. The spinning blades disappeared into the Russian's throat.

Eric relaxed. "Well, that's that. God, those damn things give me the creeps. Sometimes I think we were better off before."

"If we hadn't invented them, they would have." Leone lit a cigarette shakily. "I wonder why a Russian would come all this way alone. I didn't see anyone covering him."

Lt. Scott came slipping up the tunnel, into the bunker. "What happened? Something entered the screen."

"An Ivan."

"Just one?"

Eric brought the view screen around. Scott peered into it. Now there were numerous metal spheres crawling over the prostrate body, dull metal globes clicking and whirring, sawing up the Russian into small parts to be carried away.

"What a lot of claws," Scott murmured.

"They come like flies. Not much game for them any more."

Scott pushed the sight away, disgusted. "Like flies. I wonder why he was out there. They know we have claws all around."

A larger robot had joined the smaller spheres. It was directing operations, a long blunt tube with projecting eyepieces. There was not much left of the soldier. What remained was being brought down the hillside by the host of claws.

"Sir," Leone said. "If it's all right, I'd like to go out there and take a look at him."

"Why?"

"Maybe he came with something."

Scott considered. He shrugged. "All right. But be careful."

"I have my tab." Leone patted the metal band at his wrist. "I'll be out of bounds."

He picked up his rifle and stepped carefully up to the mouth of the bunker, making his way between blocks of concrete and steel prongs, twisted and bent. The air was cold at the top. He crossed over the ground toward the remains of the soldier, striding across the soft ash. A wind blew around him, swirling gray particles up in his face. He squinted and pushed on.

The claws retreated as he came close, some of them stiffening into immobility. He touched his tab. The Ivan would have given something for that! Short hard radiation emitted from the tab neutralized the claws, put them out of commission. Even the big robot with its two waving eyestalks retreated respectfully as he approached.

He bent down over the remains of the soldier. The gloved hand was closed tightly. There was something in it. Leone pried the fingers apart. A sealed container, aluminum. Still shiny.

He put it in his pocket and made his way back to the bunker. Behind him the claws came back to life, moving into operation again. The procession resumed, metal spheres moving through the gray ash with their loads. He could hear their treads scrabbling against the ground. He shuddered.

Scott watched intently as he brought the shiny tube out of his pocket. "He had that?"

"In his hand." Leone unscrewed the top. "Maybe you should look at it, sir."

Scott took it. He emptied the contents out in the palm of his hand. A small piece of silk paper, carefully folded. He sat down by the light and unfolded it.

"What's it say, sir?" Eric said. Several officers came up the tunnel. Major Hendricks appeared.

"Major," Scott said. "Look at this."

Hendricks read the slip. "This just come?"

"A single runner. Just now."

"Where is he?" Hendricks asked sharply.

"The claws got him."

Major Hendricks grunted. "Here." He passed it to his companions. "I think this is what we've been waiting for. They certainly took their time about it."

"So they want to talk terms," Scott said. "Are we going along with them?"

"That's not for us to decide." Hendricks sat down. "Where's the communications officer? I want the Moon Base."

Leone pondered as the communications officer raised the outside antenna cautiously, scanning the sky above the bunker for any sign of a watching Russian ship.

"Sir," Scott said to Hendricks. "It's sure strange they suddenly came around. We've been using the claws for almost a year. Now all of a sudden they start to fold."

"Maybe claws have been getting down in their bunkers."

"One of the big ones, the kind with stalks, got into an Ivan bunker last week," Eric said. "It got a whole platoon of them before they got their lid shut."

"How do you know?"

"A buddy told me. The thing came back with – with remains."

"Moon Base, sir," the communications officer said.

On the screen the face of the lunar monitor appeared. His crisp uniform contrasted to the uniforms in the bunker. And he was clean shaven. "Moon Base."

"This is forward command L-Whistle. On Terra. Let me have General Thompson."

The monitor faded. Presently General Thompson's heavy features came into focus. "What is it, Major?"

"Our claws got a single Russian runner with a message. We don't know whether to act on it – there have been tricks like this in the past."

"What's the message?"

"The Russians want us to send a single officer on policy level over to their lines. For a conference. They don't state the nature of the conference. They say that matters of – " He consulted the slip. " – Matters of grave urgency make it advisable that discussion be opened between a representative of the UN forces and themselves."

He held the message up to the screen for the general to scan. Thompson's eyes moved.

"What should we do?" Hendricks said.

"Send a man out."

"You don't think it's a trap?"

"It might be. But the location they give for their forward command is correct. It's worth a try, at any rate."

"I'll send an officer out. And report the results to you as soon as he returns."

"All right, Major." Thompson broke the connection. The screen died. Up above, the antenna came slowly down.

Hendricks rolled up the paper, deep in thought.

"I'll go," Leone said.

"They want somebody at policy level." Hendricks rubbed his jaw. "Policy level. I haven't been outside in months. Maybe I could use a little air."

"Don't you think it's risky?"

Hendricks lifted the view sight and gazed into it. The remains of the Russian were gone. Only a single claw was in sight. It was folding itself back, disappearing into the ash, like a crab. Like some hideous metal crab. ...

"That's the only thing that bothers me." Hendricks rubbed his wrist. "I know I'm safe as long as I have this on me. But there's something about them. I hate the damn things. I wish we'd never invented them. There's something wrong with them. Relentless little – "

"If we hadn't invented them, the Ivans would have."

Hendricks pushed the sight back. "Anyhow, it seems to be winning the war. I guess that's good."

"Sounds like you're getting the same jitters as the Ivans." Hendricks examined his wrist watch. "I guess I had better get started, if I want to be there before dark."

He took a deep breath and then stepped out onto the gray, rubbled ground. After a minute he lit a cigarette and stood gazing around him. The landscape was dead. Nothing stirred. He could see for miles, endless ash and slag, ruins of buildings. A few trees without leaves or branches, only the trunks. Above him the eternal rolling clouds of gray, drifting between Terra and the sun.

Major Hendricks went on. Off to the right something scuttled, something round and metallic. A claw, going lickety-split after something. Probably after a small animal, a rat. They got rats, too. As a sort of sideline.

He came to the top of the little hill and lifted his fieldglasses. The Russian lines were a few miles ahead of him. They had a forward command post there. The runner had come from it.

A squat robot with undulating arms passed by him, its arms weaving inquiringly. The robot went on its way, disappearing under some debris. Hendricks watched it go. He had never seen that type before. There were getting to be more and more types he had never seen, new varieties and sizes coming up from the underground factories.

Hendricks put out his cigarette and hurried on. It was interesting, the use of artificial forms in warfare. How had they got started? Necessity. The Soviet Union had gained great initial success, usual with the side that got the war going. Most of North America had been blasted off the map. Retaliation was quick in coming, of course. The sky was full of circling disc-bombers long before the war began; they had been up there for years. The discs began sailing down all over Russia within hours after Washington got it.

But that hadn't helped Washington.

The American bloc governments moved to the Moon Base the first year. There was not much else to do. Europe was gone; a slag heap with dark weeds growing from the ashes and bones. Most of North America was useless; nothing could be planted, no one could live. A few million people kept going up in Canada and down in South America. But during the second year Soviet parachutists began to drop, a few at first, then more and more. They wore the first really effective anti-radiation equipment; what was left of American production moved to the moon along with the governments.

All but the troops. The remaining troops stayed behind as best they could, a few thousand here, a platoon there. No one knew exactly where they were; they stayed where they could, moving around at night, hiding in ruins, in sewers, cellars, with the rats and snakes. It looked as if the Soviet Union had the war almost won. Except for a handful of projectiles fired off from the moon daily, there was almost no weapon in use against them. They came and went as they pleased. The war, for all practical purposes, was over. Nothing effective opposed them.

And then the first claws appeared. And overnight the complexion of the war changed.

The claws were awkward, at first. Slow. The Ivans knocked them off almost as fast as they crawled out of their underground tunnels. But then they got better, faster and more cunning. Factories, all on Terra, turned them out. Factories a long way under ground, behind the Soviet lines, factories that had once made atomic projectiles, now almost forgotten.

The claws got faster, and they got bigger. New types appeared, some with feelers, some that flew. There were a few jumping kinds.

The best technicians on the moon were working on designs, making them more and more intricate, more flexible. They became uncanny; the Ivans were having a lot of trouble with them. Some of the little claws were learning to hide themselves, burrowing down into the ash, lying in wait.

And then they started getting into the Russian bunkers, slipping down when the lids were raised for air and a look around. One claw inside a bunker, a churning sphere of blades and metal – that was enough. And when one got in others followed. With a weapon like that the war couldn't go on much longer.

Maybe it was already over.

Maybe he was going to hear the news. Maybe the Politburo had decided to throw in the sponge. Too bad it had taken so long. Six years. A long time for war like that, the way they had waged it. The automatic retaliation discs, spinning down all over Russia, hundreds of thousands of them. Bacteria crystals. The Soviet guided missiles, whistling through the air. The chain bombs. And now this, the robots, the claws –

The claws weren't like other weapons. They were *alive*, from any practical standpoint, whether the Governments wanted to admit it or not. They were not machines. They were living things, spinning, creeping, shaking themselves up suddenly from the gray ash and darting toward a man, climbing up him, rushing for his throat. And that was what they had been designed to do. Their job.

They did their job well. Especially lately, with the new designs coming up. Now they repaired themselves. They were on their own. Radiation tabs protected the UN troops, but if a man lost his tab he was fair game for the claws, no matter what his uniform. Down below the surface automatic machinery stamped them out. Human beings stayed a long way off. It was too risky; nobody wanted to be around them. They were left to themselves. And they seemed to be doing all right. The new designs were faster, more complex. More efficient.

Apparently they had won the war.

Major Hendricks lit a second cigarette. The landscape depressed him. Nothing but ash and ruins. He seemed to be alone, the only living thing in the whole world. To the right the ruins of a town rose up, a few walls and heaps of debris. He tossed the dead match away, increasing his pace. Suddenly he stopped, jerking up his gun, his body tense. For a minute it looked like –

From behind the shell of a ruined building a figure came, walking slowly toward him, walking hesitantly.

Hendricks blinked. "Stop!"

The boy stopped. Hendricks lowered his gun. The boy stood silently, looking at him. He was small, not very old. Perhaps eight. But it was hard to tell. Most of the kids who remained were stunted. He wore a faded blue sweater, ragged with dirt, and short pants. His hair was long and matted. Brown hair. It hung over his face and around his ears. He held something in his arms.

"What's that you have?" Hendricks said sharply.

The boy held it out. It was a toy, a bear. A teddy bear. The boy's eyes were large, but without expression.

Hendricks relaxed. "I don't want it. Keep it."

The boy hugged the bear again.

"Where do you live?" Hendricks said.

"In there."

"The ruins?"

"Yes."

"Underground?"

"Yes."

"How many are there?"

"How – how many?"

"How many of you. How big's your settlement?"

The boy did not answer.

Hendricks frowned. "You're not all by yourself, are you?"

The boy nodded.

"How do you stay alive?"

"There's food."

"What kind of food?"

"Different."

Hendricks studied him. "How old are you?"

"Thirteen."

It wasn't possible. Or was it? The boy was thin, stunted. And probably sterile. Radiation exposure, years straight. No wonder he was so small. His arms and legs were like pipecleaners, knobby, and thin. Hendricks touched the boy's arm. His skin was dry and rough; radiation skin. He bent down, looking into the boy's face. There was no expression. Big eyes, big and dark.

"Are you blind?" Hendricks said.

"No. I can see some."

"How do you get away from the claws?"

"The claws?"

"The round things. That run and burrow."

"I don't understand."

Maybe there weren't any claws around. A lot of areas were free. They collected mostly around bunkers, where there were people. The claws had been designed to sense warmth, warmth of living things.

"You're lucky." Hendricks straightened up. "Well? Which way are you going? Back – back there?"

"Can I come with you?"

"With *me*?" Hendricks folded his arms. "I'm going a long way. Miles. I have to hurry." He looked at his watch. "I have to get there by nightfall."

"I want to come."

Hendricks fumbled in his pack. "It isn't worth it. Here." He tossed down the food cans he had with him. "You take these and go back. Okay?"

The boy said nothing.

"I'll be coming back this way. In a day or so. If you're around here when I come back you can come along with me. All right?"

"I want to go with you now."

"It's a long walk."

"I can walk."

Hendricks shifted uneasily. It made too good a target, two people walking along. And the boy would slow him down. But he might not come back this way. And if the boy were really all alone –

"Okay. Come along."

The boy fell in beside him. Hendricks strode along. The boy walked silently, clutching his teddy bear.

"What's your name?" Hendricks said, after a time.

"David Edward Derring."

"David? What – what happened to your mother and father?"

"They died."

"How?"

"In the blast."

"How long ago?"

"Six years."

Hendricks slowed down. "You've been alone six years?"

"No. There were other people for awhile. They went away."

"And you've been alone since?"

"Yes."

Hendricks glanced down. The boy was strange, saying very little. Withdrawn. But that was the way they were, the children who had survived. Quiet. Stoic. A strange kind of fatalism gripped them. Nothing came as a surprise. They accepted anything that came along. There was no longer any *normal*, any natural course of things, moral or physical, for them to expect. Custom, habit, all the determining forces of learning were gone; only brute experience remained.

"Am I walking too fast?" Hendricks said.

"No."

"How did you happen to see me?"

"I was waiting."

"Waiting?" Hendricks was puzzled. "What were you waiting for?"

"To catch things."

"What kind of things?"

"Things to eat."

"Oh." Hendricks set his lips grimly. A thirteen year old boy, living on rats and gophers and half-rotten canned food. Down in a hole under the ruins of a town. With radiation pools and claws, and Russian dive-mines up above, coasting around in the sky.

"Where are we going?" David asked.

"To the Russian lines."

"Russian?"

"The enemy. The people who started the war. They dropped the first radiation bombs. They began all this."

The boy nodded. His face showed no expression.

"I'm an American," Hendricks said.

There was no comment. On they went, the two of them, Hendricks walking a little ahead, David trailing behind him, hugging his dirty teddy bear against his chest.

About four in the afternoon they stopped to eat. Hendricks built a fire in a hollow between some slabs of concrete. He cleared the weeds away and heaped up bits of wood. The Russians' lines were not very far ahead. Around him was what had once been a long valley, acres of fruit trees and grapes. Nothing remained now but a few bleak stumps and the mountains that stretched across the horizon at the far end. And the clouds of rolling ash that blew and drifted with the wind, settling over the weeds and remains of buildings, walls here and there, once in awhile what had been a road.

Hendricks made coffee and heated up some boiled mutton and bread. "Here." He handed bread and mutton to David. David squatted by the edge of the fire, his knees knobby and white. He examined the food and then passed it back, shaking his head.

"No."

"No? Don't you want any?"

"No."

Hendricks shrugged. Maybe the boy was a mutant, used to special food. It didn't matter. When he was hungry he would find something to eat. The boy was strange. But there were many strange changes coming over the world. Life was not the same, anymore. It would never be the same again. The human race was going to have to realize that.

"Suit yourself," Hendricks said. He ate the bread and mutton by himself, washing it down with coffee. He ate slowly, finding the food hard to digest. When he was done he got to his feet and stamped the fire out.

David rose slowly, watching him with his young-old eyes.

"We're going," Hendricks said.

"All right."

Hendricks walked along, his gun in his arms. They were close; he was tense, ready for anything. The Russians should be expecting a runner, an answer to their own runner, but they were tricky. There was always the possibility of a slipup. He scanned the landscape around him. Nothing but slag and ash, a few hills, charred trees. Concrete walls. But someplace ahead was the first bunker of the Russian lines, the forward command. Underground, buried deep, with only a periscope showing, a few gun muzzles. Maybe an antenna.

"Will we be there soon?" David asked.

"Yes. Getting tired?"

"No."

"Why, then?"

David did not answer. He plodded carefully along behind, picking his way over the ash. His legs and shoes were gray with dust. His pinched face was streaked, lines of gray ash in riverlets down the pale white of his skin. There was no color to his face. Typical of the new children, growing up in cellars and sewers and underground shelters.

Hendricks slowed down. He lifted his fieldglasses and studied the ground ahead of him. Were they there, someplace, waiting for him? Watching him, the way his men had watched the Russian runner? A chill went up his back. Maybe they were getting their guns ready, preparing to fire, the way his men had prepared, made ready to kill.

Hendricks stopped, wiping perspiration from his face. "Damn." It made him uneasy. But he should be expected. The situation was different.

He strode over the ash, holding his gun tightly with both hands. Behind him came David. Hendricks peered around, tight-lipped. Any second it might happen. A burst of white light, a blast, carefully aimed from inside a deep concrete bunker.

He raised his arm and waved it around in a circle.

Nothing moved. To the right a long ridge ran, topped with dead tree trunks. A few wild vines had grown up around the trees, remains of arbors. And the eternal dark weeds. Hendricks studied the ridge. Was anything up there? Perfect place for a lookout. He approached the ridge warily, David coming silently behind. If it were his command he'd have a sentry up there, watching for troops trying to infiltrate into the command area. Of course, if it were his command there would be the claws around the area for full protection.

He stopped, feet apart, hands on his hips.

"Are we there?" David said.

"Almost."

"Why have we stopped?"

"I don't want to take any chances." Hendricks advanced slowly. Now the ridge lay directly beside him, along his right. Overlooking him. His uneasy feeling increased. If an Ivan were up there he wouldn't have a chance. He waved his arm again. They should be expecting someone in the UN uniform, in response to the note capsule. Unless the whole thing was a trap.

"Keep up with me." He turned toward David. "Don't drop behind."

"With you?"

"Up beside me! We're close. We can't take any chances. Come on."

"I'll be all right." David remained behind him, in the rear, a few paces away, still clutching his teddy bear.

"Have it your way." Hendricks raised his glasses again, suddenly tense. For a moment – had something moved? He scanned the ridge carefully. Everything was silent. Dead. No life up there, only tree trunks and ash. Maybe a few rats. The big black rats that had survived the claws. Mutants – built their own shelters out of saliva and ash. Some kind of plaster. Adaptation. He started forward again.

A tall figure came out on the ridge above him, cloak flapping. Gray-green. A Russian. Behind him a second soldier appeared, another Russian. Both lifted their guns, aiming.

Hendricks froze. He opened his mouth. The soldiers were kneeling, sighting down the side of the slope. A third figure had joined them on the ridge top, a smaller figure in gray-green. A woman. She stood behind the other two.

Hendricks found his voice. "Stop!" He waved up at them frantically. "I'm – "

The two Russians fired. Behind Hendricks there was a faint *pop*. Waves of heat lapped against him, throwing him to the ground. Ash tore at his face, grinding into his eyes and nose. Choking, he pulled himself to his knees. It was all a trap. He was finished. He had come to be killed, like a steer. The soldiers and the woman were coming down the side of the ridge toward him, sliding down through the soft ash. Hendricks was numb. His head throbbed. Awkwardly, he got his rifle up and took aim. It weighed a thousand tons; he could hardly hold it. His nose and cheeks stung. The air was full of the blast smell, a bitter acrid stench.

"Don't fire," the first Russian said, in heavily accented English.

The three of them came up to him, surrounding him. "Put down your rifle, Yank," the other said.

Hendricks was dazed. Everything had happened so fast. He had been caught. And they had blasted the boy. He turned his head. David was gone. What remained of him was strewn across the ground.

The three Russians studied him curiously. Hendricks sat, wiping blood from his nose, picking out bits of ash. He shook his head, trying to clear it. "Why did you do it?" he murmured thickly. "The boy."

"Why?" One of the soldiers helped him roughly to his feet. He turned Hendricks around. "Look."

Hendricks closed his eyes.

"Look!" The two Russians pulled him forward. "See. Hurry up. There isn't much time to spare, Yank!"

Hendricks looked. And gasped.

"See now? Now do you understand?"

From the remains of David a metal wheel rolled. Relays, glinting metal. Parts, wiring. One of the Russians kicked at the heap of remains. Parts popped out, rolling away, wheels and springs and rods. A plastic section fell in, half charred. Hendricks bent shakily down. The front of the head had come off. He could make out the intricate brain, wires and relays, tiny tubes and switches, thousands of minute studs –

"A robot," the soldier holding his arm said. "We watched it tagging you."

"Tagging me?"

"That's their way. They tag along with you. Into the bunker. That's how they get in."

Hendricks blinked, dazed. "But – "

"Come on." They led him toward the ridge. "We can't stay here. It isn't safe. There must be hundreds of them all around here."

The three of them pulled him up the side of the ridge, sliding and slipping on the ash. The woman reached the top and stood waiting for them.

"The forward command," Hendricks muttered. "I came to negotiate with the Soviet – "

"There is no more forward command. *They* got in. We'll explain." They reached the top of the ridge. "We're all that's left. The three of us. The rest were down in the bunker."

"This way. Down this way." The woman unscrewed a lid, a gray manhole cover set in the ground. "Get in."

Hendricks lowered himself. The two soldiers and the woman came behind him, following him down the ladder. The woman closed the lid after them, bolting it tightly into place.

"Good thing we saw you," one of the two soldiers grunted. "It had tagged you about as far as it was going to."

"Give me one of your cigarettes," the woman said. "I haven't had an American cigarette for weeks."

Hendricks pushed the pack to her. She took a cigarette and passed the pack to the two soldiers. In the corner of the small room the lamp gleamed fitfully. The room was low-ceilinged, cramped. The four of them sat around a small wood table. A few dirty dishes were stacked to one side. Behind a ragged curtain a second room was partly visible. Hendricks saw the corner of a cot, some blankets, clothes hung on a hook.

"We were here," the soldier beside him said. He took off his helmet, pushing his blond hair back. "I'm Corporal Rudi Maxer. Polish. Impressed in the Soviet Army two years ago." He held out his hand.

Hendricks hesitated and then shook. "Major Joseph Hendricks."

"Klaus Epstein." The other soldier shook with him, a small dark man with thinning hair. Epstein plucked nervously at his ear. "Austrian. Impressed God knows when. I don't remember. The three of us were here, Rudi and I, with Tasso." He indicated the woman. "That's how we escaped. All the rest were down in the bunker."

"And – and *they* got in?"

Epstein lit a cigarette. "First just one of them. The kind that tagged you. Then it let others in."

Hendricks became alert. "The *kind*? Are there more than one kind?"

"The little boy. David. David holding his teddy bear. That's Variety Three. The most effective."

"What are the other types?"

Epstein reached into his coat. "Here." He tossed a packet of photographs onto the table, tied with a string. "Look for yourself."

Hendricks untied the string.

"You see," Rudi Maxer said, "that was why we wanted to talk terms. The Russians, I mean. We found out about a week ago. Found out that your claws were beginning to make up new designs on their own. New types of their own. Better types. Down in your underground factories behind our lines. You let them stamp themselves, repair themselves. Made them more and more intricate. It's your fault this happened."

Hendricks examined the photos. They had been snapped hurriedly; they were blurred and indistinct. The first few showed – David. David walking along a road, by himself. David and another David. Three Davids. All exactly alike. Each with a ragged teddy bear.

All pathetic.

"Look at the others," Tasso said.

The next pictures, taken at a great distance, showed a towering wounded soldier sitting by the side of a path, his arm in a sling, the stump of one leg extended, a crude crutch on his lap. Then two wounded soldiers, both the same, standing side by side.

"That's Variety One. The Wounded Soldier." Klaus reached out and took the pictures. "You see, the claws were designed to get to human beings. To find them. Each kind was better than the last. They got farther, closer, past most of our defenses, into our lines. But as long as they were merely *machines*, metal spheres with claws and horns, feelers, they could be picked off like any other object. They could be detected as lethal robots as soon as they were seen. Once we caught sight of them – "

"Variety One subverted our whole north wing," Rudi said. "It was a long time before anyone caught on. Then it was too late. They came in, wounded soldiers, knocking and begging to be let in. So we let them in. And as soon as they were in they took over. We were watching out for machines ... ."

"At that time it was thought there was only the one type," Klaus Epstein said. "No one suspected there were other types. The pictures were flashed to us. When the runner was sent to you, we knew of just one type. Variety One. The big Wounded Soldier. We thought that was all."

"Your line fell to – "

"To Variety Three. David and his bear. That worked even better." Klaus smiled bitterly. "Soldiers are suckers for children. We brought them in and tried to feed them. We found out the hard way what they were after. At least, those who were in the bunker."

"The three of us were lucky," Rudi said. "Klaus and I were – were visiting Tasso when it happened. This is her place." He waved a big hand around. "This little cellar. We finished and climbed the ladder to start back. From the ridge we saw. There they were, all around the bunker. Fighting was still going on. David and his bear. Hundreds of them. Klaus took the pictures."

Klaus tied up the photographs again.

"And it's going on all along your line?" Hendricks said.

"Yes."

"How about *our* lines?" Without thinking, he touched the tab on his arm. "Can they—"

"They're not bothered by your radiation tabs. It makes no difference to them, Russian, American, Pole, German. It's all the same. They're doing what they were designed to do. Carrying out the original idea. They track down life, wherever they find it."

"They go by warmth," Klaus said. "That was the way you constructed them from the very start. Of course, those you designed were kept back by the radiation tabs you wear. Now they've got around that. These new varieties are lead-lined."

"What's the other variety?" Hendricks asked. "The David type, the Wounded Soldier—what's the other?"

"We don't know." Klaus pointed up at the wall. On the wall were two metal plates, ragged at the edges. Hendricks got up and studied them. They were bent and dented.

"The one on the left came off a Wounded Soldier," Rudi said. "We got one of them. It was going along toward our old bunker. We got it from the ridge, the same way we got the David tagging you."

The plate was stamped: I-V. Hendricks touched the other plate. "And this came from the David type?"

"Yes." The plate was stamped: III-V.

Klaus took a look at them, leaning over Hendricks' broad shoulder. "You can see what we're up against. There's another type. Maybe it was abandoned. Maybe it didn't work. But there must be a Second Variety. There's One and Three."

"You were lucky," Rudi said. "The David tagged you all the way here and never touched you. Probably thought you'd get it into a bunker, somewhere."

"One gets in and it's all over," Klaus said. "They move fast. One lets all the rest inside. They're inflexible. Machines with one purpose. They were built for only one thing." He rubbed sweat from his lip. "We saw."

They were silent.

"Let me have another cigarette, Yank," Tasso said. "They are good. I almost forgot how they were."

It was night. The sky was black. No stars were visible through the rolling clouds of ash. Klaus lifted the lid cautiously so that Hendricks could look out.

Rudi pointed into the darkness. "Over that way are the bunkers. Where we used to be. Not over half a mile from us. It was just chance Klaus and I were not there when it happened. Weakness. Saved by our lusts."

"All the rest must be dead," Klaus said in a low voice.

"It came quickly. This morning the Politburo reached their decision. They notified us – forward command. Our runner was sent out at once. We saw him start toward the direction of your lines. We covered him until he was out of sight."

"Alex Radrivsky. We both knew him. He disappeared about six o'clock. The sun had just come up. About noon Klaus and I had an hour relief. We crept off, away from the bunkers. No one was watching. We came here. There used to be a town here, a few houses, a street. This cellar was part of a big farmhouse. We knew Tasso would be here, hiding down in her little place. We had come here before. Others from the bunkers came here. Today happened to be our turn."

"So we were saved," Klaus said. "Chance. It might have been others. We – we finished, and then we came up to the surface and started back along the ridge. That was when we saw them, the Davids. We understood right away. We had seen the photos of the First Variety, the Wounded Soldier. Our Commissar distributed them to us with an explanation. If we had gone another step they would have seen us. As it was we had to blast two Davids before we got back. There were hundreds of them, all around. Like ants. We took pictures and slipped back here, bolting the lid tight."

"They're not so much when you catch them alone. We moved faster than they did. But they're inexorable. Not like living things. They came right at us. And we blasted them."

Major Hendricks rested against the edge of the lid, adjusting his eyes to the darkness. "Is it safe to have the lid up at all?"

"If we're careful. How else can you operate your transmitter?"

Hendricks lifted the small belt transmitter slowly. He pressed it against his ear. The metal was cold and damp. He blew against the mike, raising up the short antenna. A faint hum sounded in his ear. "That's true, I suppose."

But he still hesitated.

"We'll pull you under if anything happens," Klaus said.

"Thanks." Hendricks waited a moment, resting the transmitter against his shoulder. "Interesting, isn't it?"

"What?"

"This, the new types. The new varieties of claws. We're completely at their mercy, aren't we? By now they've probably gotten into the UN lines, too. It makes me wonder if we're not seeing the beginning of a new species. *The* new species. Evolution. The race to come after man."

Rudi grunted. "There is no race after man."

"No? Why not? Maybe we're seeing it now, the end of human beings, the beginning of the new society."

"They're not a race. They're mechanical killers. You made them to destroy. That's all they can do. They're machines with a job."

"So it seems now. But how about later on? After the war is over. Maybe, when there aren't any humans to destroy, their real potentialities will begin to show."

"You talk as if they were alive!"

"Aren't they?"

There was silence. "They're machines," Rudi said. "They look like people, but they're machines."

"Use your transmitter, Major," Klaus said. "We can't stay up here forever."

Holding the transmitter tightly Hendricks called the code of the command bunker. He waited, listening. No response. Only silence. He checked the leads carefully. Everything was in place.

"Scott!" he said into the mike. "Can you hear me?"

Silence. He raised the gain up full and tried again. Only static.

"I don't get anything. They may hear me but they may not want to answer."

"Tell them it's an emergency."

"They'll think I'm being forced to call. Under your direction." He tried again, outlining briefly what he had learned. But still the phone was silent, except for the faint static.

"Radiation pools kill most transmission," Klaus said, after awhile. "Maybe that's it."

Hendricks shut the transmitter up. "No use. No answer. Radiation pools? Maybe. Or they hear me, but won't answer. Frankly, that's what I would do, if a runner tried to call from the Soviet lines. They have no reason to believe such a story. They may hear everything I say – "

"Or maybe it's too late."

Hendricks nodded.

"We better get the lid down," Rudi said nervously. "We don't want to take unnecessary chances."

They climbed slowly back down the tunnel. Klaus bolted the lid carefully into place. They descended into the kitchen. The air was heavy and close around them.

"Could they work that fast?" Hendricks said. "I left the bunker this noon. Ten hours ago. How could they move so quickly?"

ex libris

____

SLE 2019

Department of English and Philosophy
United States Military Academy
West Point, NY

MACO®

"It doesn't take them long. Not after the first one gets in. It goes wild. You know what the little claws can do. Even *one* of these is beyond belief. Razors, each finger. Maniacal."

"All right." Hendricks moved away impatiently. He stood with his back to them.

"What's the matter?" Rudi said.

"The Moon Base. God, if they've gotten there – "

"The Moon Base?"

Hendricks turned around. "They couldn't have got to the Moon Base. How would they get there? It isn't possible. I can't believe it."

"What is this Moon Base? We've heard rumors, but nothing definite. What is the actual situation? You seem concerned."

"We're supplied from the moon. The governments are there, under the lunar surface. All our people and industries. That's what keeps us going. If they should find some way of getting off Terra, onto the moon – "

"It only takes one of them. Once the first one gets in it admits the others. Hundreds of them, all alike. You should have seen them. Identical. Like ants."

"Perfect socialism," Tasso said. "The ideal of the communist state. All citizens interchangeable."

Klaus grunted angrily. "That's enough. Well? What next?"

Hendricks paced back and forth, around the small room. The air was full of smells of food and perspiration. The others watched him. Presently Tasso pushed through the curtain, into the other room. "I'm going to take a nap."

The curtain closed behind her. Rudi and Klaus sat down at the table, still watching Hendricks.

"It's up to you," Klaus said. "We don't know your situation."

Hendricks nodded.

"It's a problem." Rudi drank some coffee, filling his cup from a rusty pot. "We're safe here for awhile, but we can't stay here forever. Not enough food or supplies."

"But if we go outside – "

"If we go outside they'll get us. Or probably they'll get us. We couldn't go very far. How far is your command bunker, Major?"

"Three or four miles."

"We might make it. The four of us. Four of us could watch all sides. They couldn't slip up behind us and start tagging us. We have three rifles, three blast rifles. Tasso can have my pistol." Rudi tapped his belt. "In the Soviet army we didn't have shoes always, but we had guns. With all four of us armed one of us might get to your command bunker. Preferably you, Major."

"What if they're already there?" Klaus said.

Rudi shrugged. "Well, then we come back here."

Hendricks stopped pacing. "What do you think the chances are they're already in the American lines?"

"Hard to say. Fairly good. They're organized. They know exactly what they're doing. Once they start they go like a horde of locusts. They have to keep moving, and fast. It's secrecy and speed they depend on. Surprise. They push their way in before anyone has any idea."

"I see," Hendricks murmured.

From the other room Tasso stirred. "Major?"

Hendricks pushed the curtain back. "What?"

Tasso looked up at him lazily from the cot. "Have you any more American cigarettes left?"

Hendricks went into the room and sat down across from her, on a wood stool. He felt in his pockets. "No. All gone."

"Too bad."

"What nationality are you?" Hendricks asked after awhile.

"Russian."

"How did you get here?"

"Here?"

"This used to be France. This was part of Normandy. Did you come with the Soviet army?"

"Why?"

"Just curious." He studied her. She had taken off her coat, tossing it over the end of the cot. She was young, about twenty. Slim. Her long hair stretched out over the pillow. She was staring at him silently, her eyes dark and large.

"What's on your mind?" Tasso said.

"Nothing. How old are you?"

"Eighteen." She continued to watch him, unblinking, her arms behind her head. She had on Russian army pants and shirt. Gray-green. Thick leather belt with counter and cartridges. Medicine kit.

"You're in the Soviet army?"

"No."

"Where did you get the uniform?"

She shrugged. "It was given to me," she told him.

"How – how old were you when you came here?"

"Sixteen."

"That young?"

Her eyes narrowed. "What do you mean?"

Hendricks rubbed his jaw. "Your life would have been a lot different if there had been no war. Sixteen. You came here at sixteen. To live this way."

"I had to survive."

"I'm not moralizing."

"Your life would have been different, too," Tasso murmured. She reached down and unfastened one of her boots. She kicked the boot off, onto the floor. "Major, do you want to go in the other room? I'm sleepy."

"It's going to be a problem, the four of us here. It's going to be hard to live in these quarters. Are there just the two rooms?"

"Yes."

"How big was the cellar originally? Was it larger than this? Are there other rooms filled up with debris? We might be able to open one of them."

"Perhaps. I really don't know." Tasso loosened her belt. She made herself comfortable on the cot, unbuttoning her shirt. "You're sure you have no more cigarettes?"

"I had only the one pack."

"Too bad. Maybe if we get back to your bunker we can find some." The other boot fell. Tasso reached up for the light cord. "Good night."

"You're going to sleep?"

"That's right."

The room plunged into darkness. Hendricks got up and made his way past the curtain, into the kitchen.

And stopped, rigid.

Rudi stood against the wall, his face white and gleaming. His mouth opened and closed but no sounds came. Klaus stood in front of him, the muzzle of his pistol in Rudi's stomach. Neither of them moved. Klaus, his hand tight around his gun, his features set. Rudi, pale and silent, spread-eagled against the wall.

"What – " Hendricks muttered, but Klaus cut him off.

"Be quiet, Major. Come over here. Your gun. Get out your gun."

Hendricks drew his pistol. "What is it?"

"Cover him." Klaus motioned him forward. "Beside me. Hurry!"

Rudi moved a little, lowering his arms. He turned to Hendricks, licking his lips. The whites of his eyes shone wildly. Sweat dripped from his forehead, down his cheeks. He fixed his gaze on Hendricks. "Major, he's gone insane. Stop him." Rudi's voice was thin and hoarse, almost inaudible.

"What's going on?" Hendricks demanded.

Without lowering his pistol Klaus answered. "Major, remember our discussion? The Three Varieties? We knew about One and Three. But we didn't know about Two. At least, we didn't know before." Klaus' fingers tightened around the gun butt. "We didn't know before, but we know now."

He pressed the trigger. A burst of white heat rolled out of the gun, licking around Rudi.

"Major, this is the Second Variety."

Tasso swept the curtain aside. "Klaus! What did you do?"

Klaus turned from the charred form, gradually sinking down the wall onto the floor. "The Second Variety, Tasso. Now we know. We have all three types identified. The danger is less. I – "

Tasso stared past him at the remains of Rudi, at the blackened, smouldering fragments and bits of cloth. "You killed him."

"Him? *It*, you mean. I was watching. I had a feeling, but I wasn't sure. At least, I wasn't sure before. But this evening I was certain." Klaus rubbed his pistol butt nervously. "We're lucky. Don't you understand? Another hour and it might – "

"You were *certain*?" Tasso pushed past him and bent down, over the steaming remains on the floor. Her face became hard. "Major, see for yourself. Bones. Flesh."

Hendricks bent down beside her. The remains were human remains. Seared flesh, charred bone fragments, part of a skull. Ligaments, viscera, blood. Blood forming a pool against the wall.

"No wheels," Tasso said calmly. She straightened up. "No wheels, no parts, no relays. Not a claw. Not the Second Variety." She folded her arms. "You're going to have to be able to explain this."

Klaus sat down at the table, all the color drained suddenly from his face. He put his head in his hands and rocked back and forth.

"Snap out of it." Tasso's fingers closed over his shoulder. "Why did you do it? Why did you kill him?"

"He was frightened," Hendricks said. "All this, the whole thing, building up around us."

"Maybe."

"What, then? What do you think?"

"I think he may have had a reason for killing Rudi. A good reason."

"What reason?"

"Maybe Rudi learned something."

Hendricks studied her bleak face. "About what?" he asked.

"About him. About Klaus."

Klaus looked up quickly. "You can see what she's trying to say. She thinks I'm the Second Variety. Don't you see, Major? Now she wants you to believe I killed him on purpose. That I'm – "

"Why did you kill him, then?" Tasso said.

"I told you." Klaus shook his head wearily. "I thought he was a claw. I thought I knew."

"Why?"

"I had been watching him. I was suspicious."

"Why?"

"I thought I had seen something. Heard something. I thought I – " He stopped.

"Go on."

"We were sitting at the table. Playing cards. You two were in the other room. It was silent. I thought I heard him – *whirr*."

There was silence.

"Do you believe that?" Tasso said to Hendricks.

"Yes. I believe what he says."

"I don't. I think he killed Rudi for a good purpose." Tasso touched the rifle, resting in the corner of the room. "Major – "

"No." Hendricks shook his head. "Let's stop it right now. One is enough. We're afraid, the way he was. If we kill him we'll be doing what he did to Rudi."

Klaus looked gratefully up at him. "Thanks. I was afraid. You understand, don't you? Now she's afraid, the way I was. She wants to kill me."

"No more killing." Hendricks moved toward the end of the ladder. "I'm going above and try the transmitter once more. If I can't get them we're moving back toward my lines tomorrow morning."

Klaus rose quickly. "I'll come up with you and give you a hand."

The night air was cold. The earth was cooling off. Klaus took a deep breath, filling his lungs. He and Hendricks stepped onto the ground, out of the tunnel. Klaus planted his feet wide apart, the rifle up, watching and listening. Hendricks crouched by the tunnel mouth, tuning the small transmitter.

"Any luck?" Klaus asked presently.

"Not yet."

"Keep trying. Tell them what happened."

Hendricks kept trying. Without success. Finally he lowered the antenna. "It's useless. They can't hear me. Or they hear me and won't answer. Or – "

"Or they don't exist."

"I'll try once more." Hendricks raised the antenna. "Scott, can you hear me? Come in!"

He listened. There was only static. Then, still very faintly –

"This is Scott."

His fingers tightened. "Scott! Is it you?"

"This is Scott."

Klaus squatted down. "Is it your command?"

"Scott, listen. Do you understand? About them, the claws. Did you get my message? Did you hear me?"

"Yes." Faintly. Almost inaudible. He could hardly make out the word.

"You got my message? Is everything all right at the bunker? None of them have got in?"

"Everything is all right."

"Have they tried to get in?"

The voice was weaker.

"No."

Hendricks turned to Klaus. "They're all right."

"Have they been attacked?"

"No." Hendricks pressed the phone tighter to his ear. "Scott, I can hardly hear you. Have you notified the Moon Base? Do they know? Are they alerted?"

No answer.

"Scott! Can you hear me?"

Silence.

Hendricks relaxed, sagging. "Faded out. Must be radiation pools."

Hendricks and Klaus looked at each other. Neither of them said anything. After a time Klaus said, "Did it sound like any of your men? Could you identify the voice?"

"It was too faint."

"You couldn't be certain?"

"No."

"Then it could have been – "

"I don't know. Now I'm not sure. Let's go back down and get the lid closed."

They climbed back down the ladder slowly, into the warm cellar. Klaus bolted the lid behind them. Tasso waited for them, her face expressionless.

"Any luck?" she asked.

Neither of them answered. "Well?" Klaus said at last. "What do you think, Major? Was it your officer, or was it one of *them*?"

"I don't know."

"Then we're just where we were before."

Hendricks stared down at the floor, his jaw set. "We'll have to go. To be sure."

"Anyhow, we have food here for only a few weeks. We'd have to go up after that, in any case."

"Apparently so."

"What's wrong?" Tasso demanded. "Did you get across to your bunker? What's the matter?"

"It may have been one of my men," Hendricks said slowly. "Or it may have been one of *them*. But we'll never know standing here." He examined his watch. "Let's turn in and get some sleep. We want to be up early tomorrow."

"Early?"

"Our best chance to get through the claws should be early in the morning," Hendricks said.

The morning was crisp and clear. Major Hendricks studied the countryside through his fieldglasses.

"See anything?" Klaus said.

"No."

"Can you make out our bunkers?"

"Which way?"

"Here." Klaus took the glasses and adjusted them. "I know where to look." He looked a long time, silently.

Tasso came to the top of the tunnel and stepped up onto the ground. "Anything?"

"No." Klaus passed the glasses back to Hendricks. "They're out of sight. Come on. Let's not stay here."

The three of them made their way down the side of the ridge, sliding in the soft ash. Across a flat rock a lizard scuttled. They stopped instantly, rigid.

"What was it?" Klaus muttered.

"A lizard."

The lizard ran on, hurrying through the ash. It was exactly the same color as the ash.

"Perfect adaptation," Klaus said. "Proves we were right. Lysenko, I mean."

They reached the bottom of the ridge and stopped, standing close together, looking around them.

"Let's go." Hendricks started off. "It's a good long trip, on foot."

Klaus fell in beside him. Tasso walked behind, her pistol held alertly. "Major, I've been meaning to ask you something," Klaus said. "How did you run across the David? The one that was tagging you."

"I met it along the way. In some ruins."

"What did it say?"

"Not much. It said it was alone. By itself."

"You couldn't tell it was a machine? It talked like a living person? You never suspected?"

"It didn't say much. I noticed nothing unusual.

"It's strange, machines so much like people that you can be fooled. Almost alive. I wonder where it'll end."

"They're doing what you Yanks designed them to do," Tasso said. "You designed them to hunt out life and destroy. Human life. Wherever they find it."

Hendricks was watching Klaus intently. "Why did you ask me? What's on your mind?"

"Nothing," Klaus answered.

"Klaus thinks you're the Second Variety," Tasso said calmly, from behind them. "Now he's got his eye on you."

Klaus flushed. "Why not? We sent a runner to the Yank lines and he comes back. Maybe he thought he'd find some good game here."

Hendricks laughed harshly. "I came from the UN bunkers. There were human beings all around me."

"Maybe you saw an opportunity to get into the Soviet lines. Maybe you saw your chance. Maybe you – "

"The Soviet lines had already been taken over. Your lines had been invaded before I left my command bunker. Don't forget that."

Tasso came up beside him. "That proves nothing at all, Major."

"Why not?"

"There appears to be little communication between the varieties. Each is made in a different factory. They don't seem to work together. You might have started for the Soviet lines without knowing anything about the work of the other varieties. Or even what the other varieties were like."

"How do you know so much about the claws?" Hendricks said.

"I've seen them. I've observed them. I observed them take over the Soviet bunkers."

"You know quite a lot," Klaus said. "Actually, you saw very little. Strange that you should have been such an acute observer."

Tasso laughed. "Do you suspect me, now?"

"Forget it," Hendricks said. They walked on in silence.

"Are we going the whole way on foot?" Tasso said, after awhile. "I'm not used to walking." She gazed around at the plain of ash, stretching out on all sides of them, as far as they could see. "How dreary."

"It's like this all the way," Klaus said.

"In a way I wish you had been in your bunker when the attack came."

"Somebody else would have been with you, if not me," Klaus muttered.

Tasso laughed, putting her hands in her pockets. "I suppose so."

They walked on, keeping their eyes on the vast plain of silent ash around them.

The sun was setting. Hendricks made his way forward slowly, waving Tasso and Klaus back. Klaus squatted down, resting his gun butt against the ground.

Tasso found a concrete slab and sat down with a sigh. "It's good to rest."

"Be quiet," Klaus said sharply.

Hendricks pushed up to the top of the rise ahead of them. The same rise the Russian runner had come up, the day before. Hendricks dropped down, stretching himself out, peering through his glasses at what lay beyond.

Nothing was visible. Only ash and occasional trees. But there, not more than fifty yards ahead, was the entrance of the forward command bunker. The bunker from which he had come. Hendricks watched silently. No motion. No sign of life. Nothing stirred.

Klaus slithered up beside him. "Where is it?"

"Down there." Hendricks passed him the glasses. Clouds of ash rolled across the evening sky. The world was darkening. They had a couple of hours of light left, at the most. Probably not that much.

"I don't see anything," Klaus said.

"That tree there. The stump. By the pile of bricks. The entrance is to the right of the bricks."

"I'll have to take your word for it."

"You and Tasso cover me from here. You'll be able to sight all the way to the bunker entrance."

"You're going down alone?"

"With my wrist tab I'll be safe. The ground around the bunker is a living field of claws. They collect down in the ash. Like crabs. Without tabs you wouldn't have a chance."

"Maybe you're right."

"I'll walk slowly all the way. As soon as I know for certain – "

"If they're down inside the bunker you won't be able to get back up here. They go fast. You don't realize."

"What do you suggest?"

Klaus considered. "I don't know. Get them to come up to the surface. So you can see."

Hendricks brought his transmitter from his belt, raising the antenna. "Let's get started."

Klaus signalled to Tasso. She crawled expertly up the side of the rise to where they were sitting.

"He's going down alone," Klaus said. "We'll cover him from here. As soon as you see him start back, fire past him at once. They come quick."

"You're not very optimistic," Tasso said.

"No, I'm not."

Hendricks opened the breech of his gun, checking it carefully. "Maybe things are all right."

"You didn't see them. Hundreds of them. All the same. Pouring out like ants."

"I should be able to find out without going down all the way." Hendricks locked his gun, gripping it in one hand, the transmitter in the other. "Well, wish me luck."

Klaus put out his hand. "Don't go down until you're sure. Talk to them from up here. Make them show themselves."

Hendricks stood up. He stepped down the side of the rise.

A moment later he was walking slowly toward the pile of bricks and debris beside the dead tree stump. Toward the entrance of the forward command bunker.

Nothing stirred. He raised the transmitter, clicking it on. "Scott? Can you hear me?"

Silence.

"Scott! This is Hendricks. Can you hear me? I'm standing outside the bunker. You should be able to see me in the view sight."

He listened, the transmitter gripped tightly. No sound. Only static. He walked forward. A claw burrowed out of the ash and raced toward him. It halted a few feet away and then slunk off. A second claw appeared, one of the big ones with feelers. It moved toward him, studied him intently, and then fell in behind him, dogging respectfully after him, a few paces away. A moment later a second big claw joined it. Silently, the claws trailed him, as he walked slowly toward the bunker.

Hendricks stopped, and behind him, the claws came to a halt. He was close, now. Almost to the bunker steps.

"Scott! Can you hear me? I'm standing right above you. Outside. On the surface. Are you picking me up?"

He waited, holding his gun against his side, the transmitter tightly to his ear. Time passed. He strained to hear, but there was only silence. Silence, and faint static.

Then, distantly, metallically—

"This is Scott."

The voice was neutral. Cold. He could not identify it. But the earphone was minute.

"Scott! Listen. I'm standing right above you. I'm on the surface, looking down into the bunker entrance."

"Yes."

"Can you see me?"

"Yes."

"Through the view sight? You have the sight trained on me?"

"Yes."

Hendricks pondered. A circle of claws waited quietly around him, gray-metal bodies on all sides of him. "Is everything all right in the bunker? Nothing unusual has happened?"

"Everything is all right."

"Will you come up to the surface? I want to see you for a moment." Hendricks took a deep breath. "Come up here with me. I want to talk to you."

"Come down."

"I'm giving you an order."

Silence.

"Are you coming?" Hendricks listened. There was no response. "I order you to come to the surface."

"Come down."

Hendricks set his jaw. "Let me talk to Leone."

There was a long pause. He listened to the static. Then a voice came, hard, thin, metallic. The same as the other. "This is Leone."

"Hendricks. I'm on the surface. At the bunker entrance. I want one of you to come up here."

"Come down."

"Why come down? I'm giving you an order!"

Silence. Hendricks lowered the transmitter. He looked carefully around him. The entrance was just ahead. Almost at his feet. He lowered the antenna and fastened the transmitter to his belt. Carefully, he gripped his gun with both hands. He moved forward, a step at a time. If they could see him they knew he was starting toward the entrance. He closed his eyes a moment.

Then he put his foot on the first step that led downward.

Two Davids came up at him, their faces identical and expressionless. He blasted them into particles. More came rushing silently up, a whole pack of them. All exactly the same.

Hendricks turned and raced back, away from the bunker, back toward the rise.

At the top of the rise Tasso and Klaus were firing down. The small claws were already streaking up toward them, shining metal spheres going fast, racing frantically through the ash. But he had no time to think about that. He knelt down, aiming at the bunker entrance, gun against his cheek. The Davids were coming out in groups, clutching their teddy bears, their thin knobby legs pumping as they ran up the steps to the surface. Hendricks fired into the main body of them. They burst apart, wheels and springs flying in all directions. He fired again through the mist of particles.

A giant lumbering figure rose up in the bunker entrance, tall and swaying. Hendricks paused, amazed. A man, a soldier. With one leg, supporting himself with a crutch.

"Major!" Tasso's voice came. More firing. The huge figure moved forward, Davids swarming around it. Hendricks broke out of his freeze. The First Variety. The Wounded Soldier.

He aimed and fired. The soldier burst into bits, parts and relays flying. Now many Davids were out on the flat ground, away from the bunker. He fired again and again, moving slowly back, half-crouching and aiming.

From the rise, Klaus fired down. The side of the rise was alive with claws making their way up. Hendricks retreated toward the rise, running and crouching. Tasso had left Klaus and was circling slowly to the right, moving away from the rise.

A David slipped up toward him, its small white face expressionless, brown hair hanging down in its eyes. It bent over suddenly, opening its arms. Its teddy bear hurtled down and leaped across the ground, bounding toward him. Hendricks fired. The bear and the David both dissolved. He grinned, blinking. It was like a dream.

"Up here!" Tasso's voice. Hendricks made his way toward her. She was over by some columns of concrete, walls of a ruined building. She was firing past him, with the hand pistol Klaus had given her.

"Thanks." He joined her, grasping for breath. She pulled him back, behind the concrete, fumbling at her belt.

"Close your eyes!" She unfastened a globe from her waist. Rapidly, she unscrewed the cap, locking it into place. "Close your eyes and get down."

She threw the bomb. It sailed in an arc, an expert, rolling and bouncing to the entrance of the bunker. Two Wounded Soldiers stood uncertainly by the brick pile. More Davids poured from behind them, out onto the plain. One of the Wounded Soldiers moved toward the bomb, stooping awkwardly down to pick it up.

The bomb went off. The concussion whirled Hendricks around, throwing him on his face. A hot wind rolled over him. Dimly he saw Tasso standing behind the columns, firing slowly and methodically at the Davids coming out of the raging clouds of white fire.

Back along the rise Klaus struggled with a ring of claws circling around him. He retreated, blasting at them and moving back, trying to break through the ring.

Hendricks struggled to his feet. His head ached. He could hardly see. Everything was licking at him, raging and whirling. His right arm would not move.

Tasso pulled back toward him. "Come on. Let's go."

"Klaus – He's still up there."

"Come on!" Tasso dragged Hendricks back, away from the columns. Hendricks shook his head, trying to clear it. Tasso led him rapidly away, her eyes intense and bright, watching for claws that had escaped the blast.

One David came out of the rolling clouds of flame. Tasso blasted it. No more appeared.

"But Klaus. What about him?" Hendricks stopped, standing unsteadily. "He – "

"Come on!"

They retreated, moving farther and farther away from the bunker. A few small claws followed them for a little while and then gave up, turning back and going off.

At last Tasso stopped. "We can stop here and get our breaths."

Hendricks sat down on some heaps of debris. He wiped his neck, gasping. "We left Klaus back there."

Tasso said nothing. She opened her gun, sliding a fresh round of blast cartridges into place.

Hendricks stared at her, dazed. "You left him back there on purpose."

Tasso snapped the gun together. She studied the heaps of rubble around them, her face expressionless. As if she were watching for something.

"What is it?" Hendricks demanded. "What are you looking for? Is something coming?" He shook his head, trying to understand. What was she doing? What was she waiting for? He could see nothing. Ash lay all around them, ash and ruins. Occasional stark tree trunks, without leaves or branches. "What– "

Tasso cut him off. "Be still." Her eyes narrowed. Suddenly her gun came up. Hendricks turned, following her gaze.

Back the way they had come a figure appeared. The figure walked unsteadily toward them. Its clothes were torn. It limped as it made its way along, going very slowly and carefully. Stopping now and then, resting and getting its strength. Once it almost fell. It stood for a moment, trying to steady itself. Then it came on.

Klaus.

Hendricks stood up. "Klaus!" He started toward him. "How the hell did you– "

Tasso fired. Hendricks swung back. She fired again, the blast passing him, a searing line of heat. The beam caught Klaus in the chest. He exploded, gears and wheels flying. For a moment he continued to walk. Then he swayed back and forth. He crashed to the ground, his arms flung out. A few more wheels rolled away.

Silence.

Tasso turned to Hendricks. "Now you understand why he killed Rudi."

Hendricks sat down again slowly. He shook his head. He was numb. He could not think.

"Do you see?" Tasso said. "Do you understand?"

Hendricks said nothing. Everything was slipping away from him, faster and faster. Darkness, rolling and plucking at him.

He closed his eyes.

Hendricks opened his eyes slowly. His body ached all over. He tried to sit up but needles of pain shot through his arm and shoulder. He gasped.

"Don't try to get up," Tasso said. She bent down, putting her cold hand against his forehead.

It was night. A few stars glinted above, shining through the drifting clouds of ash. Hendricks lay back, his teeth locked. Tasso watched him impassively. She had built a fire with some wood and weeds. The fire licked feebly, hissing at a metal cup suspended over it. Everything was silent. Unmoving darkness, beyond the fire.

"So he was the Second Variety," Hendricks murmured.

"I had always thought so."

"Why didn't you destroy him sooner?" he wanted to know.

"You held me back." Tasso crossed to the fire to look into the metal cup. "Coffee. It'll be ready to drink in awhile."

She came back and sat down beside him. Presently she opened her pistol and began to disassemble the firing mechanism, studying it intently.

"This is a beautiful gun," Tasso said, half-aloud. "The construction is superb."

"What about them? The claws."

"The concussion from the bomb put most of them out of action. They're delicate. Highly organized, I suppose."

"The Davids, too?"

"Yes."

"How did you happen to have a bomb like that?"

Tasso shrugged. "We designed it. You shouldn't underestimate our technology, Major. Without such a bomb you and I would no longer exist."

"Very useful."

Tasso stretched out her legs, warming her feet in the heat of the fire. "It surprised me that you did not seem to understand, after he killed Rudi. Why did you think he – "

"I told you. I thought he was afraid."

"Really? You know, Major, for a little while I suspected you. Because you wouldn't let me kill him. I thought you might be protecting him." She laughed.

"Are we safe here?" Hendricks asked presently.

"For awhile. Until they get reinforcements from some other area." Tasso began to clean the interior of the gun with a bit of rag. She finished and pushed the mechanism back into place. She closed the gun, running her finger along the barrel.

"We were lucky," Hendricks murmured.

"Yes. Very lucky."

"Thanks for pulling me away."

Tasso did not answer. She glanced up at him, her eyes bright in the fire light. Hendricks examined his arm. He could not move his fingers. His whole side seemed numb. Down inside him was a dull steady ache.

"How do you feel?" Tasso asked.

"My arm is damaged."

"Anything else?"

"Internal injuries."

"You didn't get down when the bomb went off."

Hendricks said nothing. He watched Tasso pour the coffee from the cup into a flat metal pan. She brought it over to him.

"Thanks." He struggled up enough to drink. It was hard to swallow. His insides turned over and he pushed the pan away. "That's all I can drink now."

Tasso drank the rest. Time passed. The clouds of ash moved across the dark sky above them. Hendricks rested, his mind blank. After awhile he became aware that Tasso was standing over him, gazing down at him.

"What is it?" he murmured.

"Do you feel any better?"

"Some."

"You know, Major, if I hadn't dragged you away they would have got you. You would be dead. Like Rudi."

"I know."

"Do you want to know why I brought you out? I could have left you. I could have left you there."

"Why did you bring me out?"

"Because we have to get away from here." Tasso stirred the fire with a stick, peering calmly down into it. "No human being can live here. When their reinforcements come we won't have a chance. I've pondered about it while you were unconscious. We have perhaps three hours before they come."

"And you expect me to get us away?"

"That's right. I expect you to get us out of here."

"Why me?"

"Because I don't know any way." Her eyes shone at him in the half-light, bright and steady. "If you can't get us out of here they'll kill us within three hours. I see nothing else ahead. Well, Major? What are you going to do? I've been waiting all night. While you were unconscious I sat here, waiting and listening. It's almost dawn. The night is almost over."

Hendricks considered. "It's curious," he said at last.

"Curious?"

"That you should think I can get us out of here. I wonder what you think I can do."

"Can you get us to the Moon Base?"

"The Moon Base? How?"

"There must be some way."

Hendricks shook his head. "No. There's no way that I know of."

Tasso said nothing. For a moment her steady gaze wavered. She ducked her head, turning abruptly away. She scrambled to her feet. "More coffee?"

"No."

"Suit yourself." Tasso drank silently. He could not see her face. He lay back against the ground, deep in thought, trying to concentrate. It was hard to think. His head still hurt. And the numbing daze still hung over him.

"There might be one way," he said suddenly.

"Oh?"

"How soon is dawn?"

"Two hours. The sun will be coming up shortly."

"There's supposed to be a ship near here. I've never seen it. But I know it exists."

"What kind of a ship?" Her voice was sharp.

"A rocket cruiser."

"Will it take us off? To the Moon Base?"

"It's supposed to. In case of emergency." He rubbed his forehead.

"What's wrong?"

"My head. It's hard to think. I can hardly – hardly concentrate. The bomb."

"Is the ship near here?" Tasso slid over beside him, settling down on her haunches. "How far is it? Where is it?"

"I'm trying to think."

Her fingers dug into his arm. "Nearby?" Her voice was like iron. "Where would it be? Would they store it underground? Hidden underground?"

"Yes. In a storage locker."

"How do we find it? Is it marked? Is there a code marker to identify it?"

Hendricks concentrated. "No. No markings. No code symbol."

"What, then?"

"A sign."

"What sort of sign?"

Hendricks did not answer. In the flickering light his eyes were glazed, two sightless orbs. Tasso's fingers dug into his arm.

"What sort of sign? What is it?"

"I – I can't think. Let me rest."

"All right." She let go and stood up. Hendricks lay back against the ground, his eyes closed. Tasso walked away from him, her hands in her pockets. She kicked a rock out of her way and stood staring up at the sky. The night blackness was already beginning to fade into gray. Morning was coming.

Tasso gripped her pistol and walked around the fire in a circle, back and forth. On the ground Major Hendricks lay, his eyes closed, unmoving. The grayness rose in the sky, higher and higher. The landscape became visible, fields of ash stretching out in all directions. Ash and ruins of buildings, a wall here and there, heaps of concrete, the naked trunk of a tree.

The air was cold and sharp. Somewhere a long way off a bird made a few bleak sounds.

Hendricks stirred. He opened his eyes. "Is it dawn? Already?"

"Yes."

Hendricks sat up a little. "You wanted to know something. You were asking me."

"Do you remember now?"

"Yes."

"What is it?" She tensed. "What?" she repeated sharply.

"A well. A ruined well. It's in a storage locker under a well."

"A well." Tasso relaxed. "Then we'll find a well." She looked at her watch. "We have about an hour, Major. Do you think we can find it in an hour?"

"Give me a hand up," Hendricks said.

Tasso put her pistol away and helped him to his feet. "This is going to be difficult."

"Yes it is." Hendricks set his lips tightly. "I don't think we're going to go very far."

They began to walk. The early sun cast a little warmth down on them. The land was flat and barren, stretching out gray and lifeless as far as they could see. A few birds sailed silently, far above them, circling slowly.

"See anything?" Hendricks said. "Any claws?"

"No. Not yet."

They passed through some ruins, upright concrete and bricks. A cement foundation. Rats scuttled away. Tasso jumped back warily.

"This used to be a town," Hendricks said. "A village. Provincial village. This was all grape country, once. Where we are now."

They came onto a ruined street, weeds and cracks criss-crossing it. Over to the right a stone chimney stuck up.

"Be careful," he warned her.

A pit yawned, an open basement. Ragged ends of pipes jutted up, twisted and bent. They passed part of a house, a bathtub turned on its side. A broken chair. A few spoons and bits of china dishes. In the center of the street the ground had sunk away. The depression was filled with weeds and debris and bones.

"Over here," Hendricks murmured.

"This way?"

"To the right."

They passed the remains of a heavy duty tank. Hendricks' belt counter clicked ominously. The tank had been radiation blasted. A few feet from the tank a mummified body lay sprawled out, mouth open. Beyond the road was a flat field. Stones and weeds, and bits of broken glass.

"There," Hendricks said.

A stone well jutted up, sagging and broken. A few boards lay across it. Most of the well had sunk into rubble. Hendricks walked unsteadily toward it, Tasso beside him.

"Are you certain about this?" Tasso said. "This doesn't look like anything."

"I'm sure." Hendricks sat down at the edge of the well, his teeth locked. His breath came quickly. He wiped perspiration from his face. "This was arranged so the senior command officer could get away. If anything happened. If the bunker fell."

"That was you?"

"Yes."

"Where is the ship? Is it here?"

"We're standing on it." Hendricks ran his hands over the surface of the well stones. "The eye-lock responds to me, not to anybody else. It's my ship. Or it was supposed to be."

There was a sharp click. Presently they heard a low grating sound from below them.

"Step back," Hendricks said. He and Tasso moved away from the well.

A section of the ground slid back. A metal frame pushed slowly up through the ash, shoving bricks and weeds out of the way. The action ceased, as the ship nosed into view.

"There it is," Hendricks said.

The ship was small. It rested quietly, suspended in its mesh frame, like a blunt needle. A rain of ash sifted down into the dark cavity from which the ship had been raised. Hendricks made his way over to it. He mounted the mesh and unscrewed the hatch, pulling it back. Inside the ship the control banks and the pressure seat were visible.

Tasso came and stood beside him, gazing into the ship. "I'm not accustomed to rocket piloting," she said, after awhile.

Hendricks glanced at her. "I'll do the piloting."

"Will you? There's only one seat, Major. I can see it's built to carry only a single person."

Hendricks' breathing changed. He studied the interior of the ship intently. Tasso was right. There was only one seat. The ship was built to carry only one person. "I see," he said slowly. "And the one person is you."

She nodded.

"Of course."

"Why?"

"*You* can't go. You might not live through the trip. You're injured. You probably wouldn't get there."

"An interesting point. But you see, I know where the Moon Base is. And you don't. You might fly around for months and not find it. It's well hidden. Without knowing what to look for – "

"I'll have to take my chances. Maybe I won't find it. Not by myself. But I think you'll give me all the information I need. Your life depends on it."

"How?"

"If I find the Moon Base in time, perhaps I can get them to send a ship back to pick you up. *If* I find the Base in time. If not, then you haven't a chance. I imagine there are supplies on the ship. They will last me long enough – "

Hendricks moved quickly. But his injured arm betrayed him. Tasso ducked, sliding lithely aside. Her hand came up, lightning fast. Hendricks saw the gun butt coming. He tried to ward off the blow, but she was too fast. The metal butt struck against the side of his head, just above his ear. Numbing pain rushed through him. Pain and rolling clouds of blackness. He sank down, sliding to the ground.

Dimly, he was aware that Tasso was standing over him, kicking him with her toe.

"Major! Wake up."

He opened his eyes, groaning.

"Listen to me." She bent down, the gun pointed at his face. "I have to hurry. There isn't much time left. The ship is ready to go, but you must tell me the information I need before I leave."

Hendricks shook his head, trying to clear it.

"Hurry up! Where is the Moon Base? How do I find it? What do I look for?"

Hendricks said nothing.

"Answer me!"

"Sorry."

"Major, the ship is loaded with provisions. I can coast for weeks. I'll find the Base eventually. And in a half hour you'll be dead. Your only chance of survival – " She broke off.

Along the slope, by some crumbling ruins, something moved. Something in the ash. Tasso turned quickly, aiming. She fired. A puff of flame leaped. Something scuttled away, rolling across the ash. She fired again. The claw burst apart, wheels flying.

"See?" Tasso said. "A scout. It won't be long."

"You'll bring them back here to get me?"

"Yes. As soon as possible."

Hendricks looked up at her. He studied her intently. "You're telling the truth?" A strange expression had come over his face, an avid hunger. "You will come back for me? You'll get me to the Moon Base?"

"I'll get you to the Moon Base. But tell me where it is! There's only a little time left."

"All right." Hendricks picked up a piece of rock, pulling himself to a sitting position. "Watch."

Hendricks began to scratch in the ash. Tasso stood by him, watching the motion of the rock. Hendricks was sketching a crude lunar map.

"This is the Appenine range. Here is the Crater of Archimedes. The Moon Base is beyond the end of the Appenine, about two hundred miles. I don't know exactly where. No one on Terra knows. But when you're over the Appenine, signal with one red flare and a green flare, followed by two red flares in quick succession. The Base monitor will record your signal. The Base is under the surface, of course. They'll guide you down with magnetic grapples."

"And the controls? Can I operate them?"

"The controls are virtually automatic. All you have to do is give the right signal at the right time."

"I will."

"The seat absorbs most of the take-off shock. Air and temperature are automatically controlled. The ship will leave Terra and pass out into free space. It'll line itself up with the moon, falling into an orbit around it, about a hundred miles above the surface. The orbit will carry you over the Base. When you're in the region of the Appenine, release the signal rockets."

Tasso slid into the ship and lowered herself into the pressure seat. The arm locks folded automatically around her. She fingered the controls. "Too bad you're not going, Major. All this put here for you, and you can't make the trip."

"Leave me the pistol."

Tasso pulled the pistol from her belt. She held it in her hand, weighing it thoughtfully. "Don't go too far from this location. It'll be hard to find you, as it is."

"No. I'll stay here by the well."

Tasso gripped the take-off switch, running her fingers over the smooth metal. "A beautiful ship, Major. Well built. I admire your workmanship. You people have always done good work. You build fine things. Your work, your creations, are your greatest achievement."

"Give me the pistol," Hendricks said impatiently, holding out his hand. He struggled to his feet.

"Good-bye, Major." Tasso tossed the pistol past Hendricks. The pistol clattered against the ground, bouncing and rolling away. Hendricks hurried after it. He bent down, snatching it up.

The hatch of the ship clanged shut. The bolts fell into place. Hendricks made his way back. The inner door was being sealed. He raised the pistol unsteadily.

There was a shattering roar. The ship burst up from its metal cage, fusing the mesh behind it. Hendricks cringed, pulling back. The ship shot up into the rolling clouds of ash, disappearing into the sky.

Hendricks stood watching a long time, until even the streamer had dissipated. Nothing stirred. The morning air was chill and silent. He began to walk aimlessly back the way they had come. Better to keep moving around. It would be a long time before help came – if it came at all.

He searched his pockets until he found a package of cigarettes. He lit one grimly. They had all wanted cigarettes from him. But cigarettes were scarce.

A lizard slithered by him, through the ash. He halted, rigid. The lizard disappeared. Above, the sun rose higher in the sky. Some flies landed on a flat rock to one side of him. Hendricks kicked at them with his foot.

It was getting hot. Sweat trickled down his face, into his collar. His mouth was dry.

Presently he stopped walking and sat down on some debris. He unfastened his medicine kit and swallowed a few narcotic capsules. He looked around him. Where was he?

Something lay ahead. Stretched out on the ground. Silent and unmoving.

Hendricks drew his gun quickly. It looked like a man. Then he remembered. It was the remains of Klaus. The Second Variety. Where Tasso had blasted him. He could see wheels and relays and metal parts, strewn around on the ash. Glittering and sparkling in the sunlight.

Hendricks got to his feet and walked over. He nudged the inert form with his foot, turning it over a little. He could see the metal hull, the aluminum ribs and struts. More wiring fell out. Like viscera. Heaps of wiring, switches and relays. Endless motors and rods.

He bent down. The brain cage had been smashed by the fall. The artificial brain was visible. He gazed at it. A maze of circuits. Miniature tubes. Wires as fine as hair. He touched the brain cage. It swung aside. The type plate was visible. Hendricks studied the plate.

And blanched.

IV–IV.

For a long time he stared at the plate. Fourth Variety. Not the Second. They had been wrong. There were more types. Not just three. Many more, perhaps. At least four. And Klaus wasn't the Second Variety.

But if Klaus wasn't the Second Variety–

Suddenly he tensed. Something was coming, walking through the ash beyond the hill. What was it? He strained to see. Figures. Figures coming slowly along, making their way through the ash.

Coming toward him.

Hendricks crouched quickly, raising his gun. Sweat dripped down into his eyes. He fought down rising panic, as the figures neared.

The first was a David. The David saw him and increased its pace. The others hurried behind it. A second David. A third. Three Davids, all alike, coming toward him silently, without expression, their thin legs rising and falling. Clutching their teddy bears.

He aimed and fired. The first two Davids dissolved into particles. The third came on. And the figure behind it. Climbing silently toward him across the gray ash. A Wounded Soldier, towering over the David. And–

And behind the Wounded Soldier came two Tassos, walking side by side. Heavy belt, Russian army pants, shirt, long hair. The familiar figure, as he had seen her only a little while before. Sitting in the pressure seat of the ship. Two slim, silent figures, both identical.

They were very near. The David bent down suddenly, dropping its teddy bear. The bear raced across the ground. Automatically, Hendricks' fingers tightened around the trigger. The bear was gone, dissolved into mist. The two Tasso Types moved on, expressionless, walking side by side, through the gray ash.

When they were almost to him, Hendricks raised the pistol waist high and fired.

The two Tassos dissolved. But already a new group was starting up the rise, five or six Tassos, all identical, a line of them coming rapidly toward him.

And he had given her the ship and the signal code. Because of him she was on her way to the moon, to the Moon Base. He had made it possible.

He had been right about the bomb, after all. It had been designed with knowledge of the other types, the David Type and the Wounded Soldier Type. And the Klaus Type. Not designed by human beings. It had been designed by one of the underground factories, apart from all human contact.

The line of Tassos came up to him. Hendricks braced himself, watching them calmly. The familiar face, the belt, the heavy shirt, the bomb carefully in place.

The bomb –

As the Tassos reached for him, a last ironic thought drifted through Hendricks' mind. He felt a little better, thinking about it. The bomb. Made by the Second Variety to destroy the other varieties. Made for that end alone.

They were already beginning to design weapons to use against each other.

After You Have Read the Story ...

Historically, "Second Variety" may be the first science fiction story to feature explicitly weaponized robots and the first to combine weaponized robots with self-replicating machines. The general theme of the story, that we create something that is cleverer than us and kills us, was not novel—that theme was introduced in Karel Čapek's play *R.U.R.*, which was the origin of the term "robot."[4] Other stories before "Second Variety" explore a similar theme of how our servants become our masters, such as Jack Williamson's 1947 story "With Folded Hands," but aren't about the militarization of robots.[5] The concept of self-replicating machines was also not novel, as it appeared in science fiction as a central concept as early as 1943 with A. E. van Vogt's "M33 in Andromeda."[6]

"Second Variety" is, however, the root meme for every story, play, or movie in which a robot has masqueraded as a human in order to kill humans. Although the robot Maria in the 1927 movie *Metropolis* masqueraded as a human to manipulate workers, and her masters hoped she would kill rebelling workers, Maria did not directly attack anyone. True killer robots came later: Ash in *Alien* (1979), the terminator in the *Terminator* (1984), and the Cylons in *Battlestar Galactica* (2004–2009) all owe Dick a nod of acknowledgement.

The story provides an opportunity to explore ethics but also, as discussed below, whether machine learning can exceed bounded rationality and really produce unintended consequences. For further reading, bounded rationality is introduced in chapter 3, machine learning in chapter 16, and ethics in chapter 20 of *Introduction to AI Robotics*.[7]

When Is a Robot a Robot?

Though ethics has generally focused on how robots would treat us, it also considers how we would treat them. The previous story, "Supertoys Last All Summer Long," raises this question of how will we treat sentient robots, assuming we notice or acknowledge that they have sentience. Philosophers have begun to formally explore this aspect of robots and ethics.[8] In science fiction, the question of the moral treatment of a sentient weaponized robot is brilliantly explored in *He, She and It* by Marge Piercy.[9] In that novel, a robot has been engineered to be the equivalent of a one-person, enhanced special operations force in order to protect a small enclave in a

Mad Max–style postapocalyptic world. The problem is that the robot is a pacifist and does not want to fight. Does the robot have a right to choose not to fulfill the role it was explicitly built for?

One confusing aspect in the real-life discussion of the ethics of weaponizing robots is the question, when is a robot a robot? For example, once it is launched, a cruise missile is a physically situated agent that constantly optimizes its performance in order to meet its goal. The missile can't be recalled and soldiers may risk their lives to deploy and protect the missile. But we don't consider a cruise missile to be a robot. There is no current controversy over it, there are no ongoing United Nations commissions meeting to discuss its ramifications, and no public statements from famous scientists like Stephen Hawking. Like cruise missiles, automatic target systems on missiles and fighter aircraft are not considered robots. Still, in 2006, South Korea introduced the Samsung SGR-A1, a machine gun fixed to the ground that incorporates heat and movement tracking for guarding the demilitarized zone (DMZ), and it was called a robot.[10] It was labeled this way despite normally being under human control (though operators can switch to an automatic shoot mode, presumably if the DMZ is being overrun[11]). It also acts much like a motion-sensing light; the SGR-A1s are pointed at an area designated as a kill zone wherein anything that moves is acceptable to shoot. There is a suspicion among robot researchers, however, that labeling the machine gun a "robot" is a marketing decision intended to make the system sound smarter and more formidable.

The robots in "Second Variety" are certainly intelligent and arguably commit war crimes, leading to the issue of who would be considered responsible. The story implies that the robots begin to spontaneously create new varieties of killer robots due to a confluence of three things: an objective to kill all living things, a lack of supervision of the factories by engineers on the moon, and machine learning. The designers appear to have intended a system much like a mobile version of a Samsung SGR-A1, but giving the robots the objective of killing *all* living things without enforceable geofencing (creating a boundary with geospatial or GPS coordinates) is a war crime. Relying on a radiation tag to protect "good" people is another design flaw that leads to a war crime. But would the designers be responsible for the robots' transformation from reactive claws to deliberative humanoids? In an article on robot ethics, David Woods and I argue that the lack of accountability in creating reliable robots and ensuring correct operation without unintended consequences is a more subtle and fundamental danger than robots making unethical decisions.[12] This lack of accountability for the reliable and safe operation of weaponized robots is amplified by the current lack of adequate testing and evaluation methods, which was addressed in our discussion of "Catch That Rabbit."

Bounded Rationality and Machine Learning

Does machine learning obviate the culpability of robot designers? In science fiction, and by the media in real life, machine learning is treated like a biological virus. Learning is apparently going to enable a robot to spontaneously mutate, propagate, and exceed its programmed level of initiative (see chapter 4), which will lead to our destruction. There are two practical aspects of artificial intelligence that explain why machine learning is unlikely to cause robots to evolve to peer-level intelligence: bounded rationality and the limitations of machine learning in practice.

Herb Simon, one of the founders of the field of artificial intelligence, popularized the concept of bounded rationality, where an agent such as a human makes rational decisions within an environment that exceeds its computational ability to adapt and thus bounds what it can do.[13] In economics, people may make rational decisions about incurring debt or giving up smoking (due to cost) given what they know or can perceive, but these are not necessarily optimal decisions because they can't know or perceive everything. People and the market are bounded or finite. In artificial intelligence, bounded rationality means a computational agent is not able to exceed its resources; essentially it cannot compute "outside the box." A robot may be nondeterministic in how it avoids an obstacle, going unpredictably either left or right, but it won't jump over the obstacle if it isn't capable of jumping or hasn't been programmed to experiment with alternative motions. One question that "Second Variety" raises, assuming that its robots have been programmed with a high degree of initiative (see chapter 4), is where do the claws get the resources to add compute-intensive deliberative capabilities that go beyond their basic reactive behaviors?

The story offers learning as the mechanism by which the robots mutate into a human-like intelligence pursuing a simplistic, hostile objective.

Could learning actually allow a behavioral robot to escape computational bounds and produce deliberative capabilities? Probably not, for at least two reasons. First, even if a claw had excess compute power, it would still need some resources in the form of a favorable knowledge structure to enable learning. A learning algorithm has to be designed to improve a particular component, have some sort of representation, and receive either direct or indirect feedback. Thus an algorithm will only be as good as the computational bounds placed on it by its representation and the type of feedback it can receive about its actions. The second reason why learning would be unlikely to allow the claws to mutate is the *new term* problem. This problem refers to the difficulty a learning algorithm has in learning something truly new. For example, an algorithm for learning to classify colored objects needs to know how many different colors it is supposed to learn. The accepted wisdom is that robots can only learn what we specify and give them the resources to learn.

Learning as the magic ingredient is also problematic because there is not a single mechanism and none of the mechanisms, either individually or collectively, seem to lead to the sentience exhibited by the second variety. AI divides machine learning into four general classes: *supervised, unsupervised, semi-supervised,* and *reinforcement learning.*[14] Supervised learning is often used to train a robot to perform a task by watching how a human does it; in this case supervision refers to the robot having a source of feedback or measure of error during training. Induction, decision trees, and support vector machines are common algorithms for supervised learning. In "Second Variety," supervised learning might explain how the robots have duplicated humans by observing humans. Unsupervised learning is often used in data mining, where a robot might notice patterns. Artificial neural networks are a popular form of unsupervised learning, especially a variant known as deep convolutional neural networks or *deep learning*. There is no obvious example of where unsupervised learning might be employed by the robots in "Second Variety." Semi-supervised learning is, as the name implies, a hybrid and is fairly rare in robotics. Perhaps the most common type of learning in robotics practice is reinforcement learning. The term "reinforcement" refers to feedback from trial and error rather than negative punishment. Q-learning is a common form of reinforcement learning and requires a robot to explore many possible options to achieve a specific objective, often in simulation, in order to learn policies of what to do for various situations. The claws learning to hide would likely have been accomplished with a reinforcement learning algorithm.

Of all the learning algorithms, artificial neural networks seems to be the one most bandied about by science fiction writers. There may be at least three reasons for this. One is that neural networks is an unsupervised type of learning and thus sounds like it can discover new knowledge. The newness of what a neural net learns is debatable, however, given that we tell it the features or symbols to look for as part of the hidden structure of an object. If we give a robot many examples of a red ball to follow and the ability to extract "red" and "round," then it will converge on identifying red balls in images and may even weakly identify round things and red things. But it hasn't really learned a new idea, or solved the new term problem in AI, as it was given all the pieces it needed and thus bounded in what it could do. The second reason why neural networks may be a favorite enabler for robot apocalypses is that the algorithm produces a classifier and it is often difficult to explicitly express why it works. This is particularly true for object recognition where a neural network can often recognize an object that is smaller or larger than what it has been shown, is shadowed in the dark, or is partially hidden. The network is learning a very dense, weighted, nonlinear relationship between many features that is extremely hard to translate into a form that humans can understand. As a result, there is a bit of mystery about neural networks and thus it is easy to imagine that unbounded relationships are being created. The third reason is that deep learning sounds, well … deep and thoughtful and ponderous. If the robot is deep in thought, it is bound to learn something new, right? Not necessarily, as it is "deep" only in terms of layers of dense statistical relationships between features, and not in terms of profundity.

Regardless of feasibility, "Second Variety" sums up pervasive apprehensions about machine intelligence, especially those that say we will apply it unwisely and that it will get away from our control. Nothing in fifty years of AI research suggests that robots can exceed their level of initiative and the bounds on their rationality imposed through design. The real question is, will we apply robotics unwisely? The robots aren't responsible for the horrible state of the world in "Second Variety"; the responsible party is the nation that produced the robots and has accepted criminal forms of lethality. Perhaps the real apprehension over weaponized robots is that we will design and program what we deserve.

Reality Score: C–

The only realistic aspect of "Second Variety" is the claw robots and their behavior-based operation; the machine learning and self-replication is fanciful at best and embarrassingly wrong at worst. The story inevitably leads to a discussion of ethical issues but does not address ethics itself.

7 Summary and Review

The six stories we've discussed in this book are remarkable for their vision of robots and for foreseeing the actual science that might make robots inarguably intelligent. That the stories have some accurate elements is even more remarkable considering that five of the six were published before the first autonomous mobile robot, Shakey, was completed in 1967.[1] The only story in this collection that came *after* Shakey was "Stranger in Paradise," which was about telepresence, not autonomy.

The stories are also noteworthy because they highlight the economic justification for robotics. In factory automation and military applications, a common categorization of types of tasks that robots would be better at—and thus more economical—than humans are called the *Three D's*: *dirty*, *dangerous*, and *dull*. Certainly the mining operations in "Catch That Rabbit" and the war in "Second Variety" represent vocations where the work is dirty, dangerous, and/or dull. Still, the Three D's often mislead designers into focusing on tasks where the robot is replacing a human in a job that already exists, which is not necessarily the case. "Stranger in Paradise," "Runaround," and "Long Shot" give examples of robots doing things that are beyond the Three D's—tasks that are simply impossible for a human. The robots in these stories either exceed the abilities of a human to work in a particular environment at all or to work persistently over the expected long time scale. These types of tasks are referred to as situations in which the robots are *better than bio*. The one exception to the Three D's or better-than-bio economic drivers is "Supertoys Last All Summer Long." In that story, the robot is literally a substitute for a human when price is no object.

The stories, by virtue of being short stories, are only able to touch on a few of the concepts in artificial intelligence for robotics. This gives a piecemeal view of intelligence, so it is useful to review the stories together and summarize the answers to the larger questions posed in the introduction of *How are intelligent robots programmed?* and especially *What are the limits of autonomous robots?* The sections below revisit these two questions. The stories and commentary do not directly address two major memes in the public's perception of autonomy: Asimov's Three Laws of Robotics and the singularity. Given that the stories and commentary have established an informed background on intelligent robots, these memes are included in the summary section on the limits of autonomous robots.

How Are Intelligent Robots Programmed?

The stories and commentary indirectly support the adage, "if it is easy for a person, it will be hard for a robot; if it is easy for a robot, it is hard for a person." The way we think we work is not the way robots work.

To recap, artificial intelligence for robots consists of four building blocks or primitives: SENSE, ACT, PLAN, LEARN. The software architecture for organizing these primitives leads to three distinct flavors of intelligence, each with a unique set of algorithms. The reactive flavor, or tier, of intelligence mimics reactive animal behaviors that use combinations of SENSE and ACT. Deliberative intelligence makes use of PLAN, SENSE, and ACT, though most often the robot PLANs infrequently to generate the SENSE and ACT behaviors, selects the best set of resources for the behaviors and environment, and then monitors for the successful execution of the plan or opportunities to improve. Many people think they SENSE, then PLAN, and then finally ACT in a continuous cycle but cognitive science has shown that people both PLAN and then SENSE-ACT and also SENSE, PLAN, and ACT.[2] Interactive, or social, intelligence applies both reactive and deliberative combinations of the primitives to interact with others, be they other robots or humans. LEARN is sprinkled throughout all three tiers, with different types of learning algorithms applied to different objectives.

Figure 7.1 provides a canonical software operational architecture that brings together the behavior-based principles discussed with "Runaround" and the deliberation and time horizons discussed with "Long Shot." The tiers within the software organizational architecture imply, incorrectly, that a programmer would start with robots that are purely reactive, then add

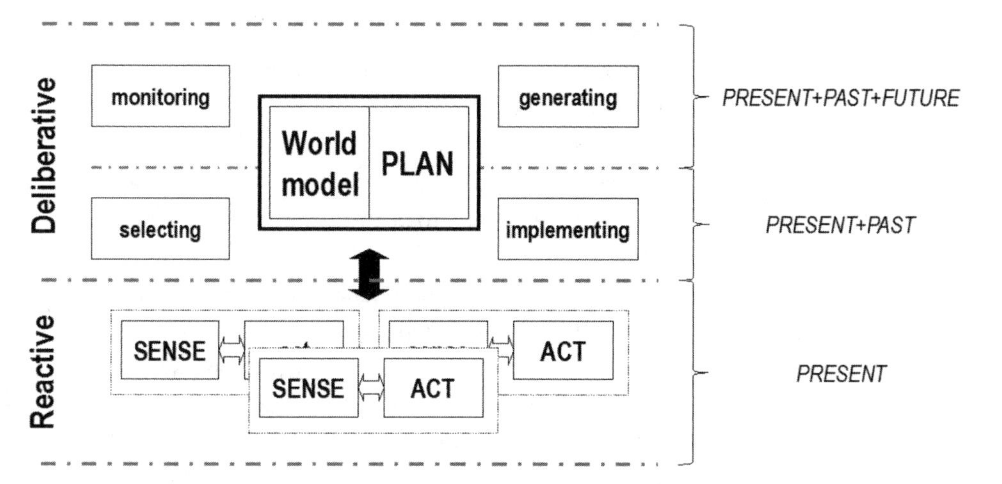

Figure 7.1
Canonical software operational architecture (from Murphy, *Introduction to AI Robotics*).

on modules to make the robot more deliberative, and then once that was done add interactive abilities and voilà! an intelligent robot. In practice, a robot has to have the right amount of reaction, deliberation, and interaction skills to successfully perform its intended tasks in the larger sociotechnical system. A robot with deliberative intelligence is not necessarily more intelligent than a robot with only reactive intelligence, and neither robot may be used if it has no interactive intelligence. The tiers help clarify what a programmer needs to take from the libraries of reactive, deliberative, and interactive intelligences in order to make their robot a success.

The dramatic tension in each story is created or heightened by the reader's perception of the literal-mindedness of software programs and bounds on the robot's computational abilities. Telepresence in "Stranger in Paradise" is required because computers in that world can't do what people can do, which was reasonably representative of the state of the practice at the time it was written. "Runaround" and "Catch That Rabbit" are amusing versions of literal-mindedness combined with the ambiguity of natural language and inability to monitor for conflicts and progress, but they do not match the current state of the practice. "Second Variety" gives a horrific version of taking defense to the extreme by playing to the reader's fears, but this vision of unbounded machine learning for a very narrow, single-minded purpose is unrealistic. The source of tension in "Long Shot," the bounds on Ilse's memory, would also be a source of tension in today's robots; Ilse's mission is decomposed into subgoals and she (and the reader) cannot see the ultimate objective until she gets to the last step. David's sentience in "Supertoys Last All Summer Long" is hidden in plain sight because the reader is not expecting a robot to be more than a literal-minded automaton.

Dramatic tension in these stories is also generated by the inability to infer why a robot is acting the way it is. One barrier to inferring why a robot is doing something is that a human character may not have a correct *theory of mind* model of what the robot is doing or how it works. The value of modeling a theory of mind for diagnosing a robot failure is debatable, however. A fundamental principle in human-computer interaction is the need to make key operations transparent. For example, when a computer is downloading, there is an hourglass or spinning wheel to indicate that the process is executing normally. Advanced users can pull up execution monitors as needed for more sophisticated dialog. There should always be diagnostic routines that can be run and always a backdoor. Donovan and Powell would not have a job if the robots in "Runaround" and "Catch That Rabbit" followed lessons learned in basic human–computer design. Monica would have a harder time ignoring David if diagnostics told her and her husband that he didn't have a problem with natural language. Only Ilse in "Long Shot" exhibits best practices and allows her designers, at least in the initial stages of her existence, to directly observe what is going on inside her.

What Are the Limits of Autonomous Robots?

All the stories except for "Stranger in Paradise" are about robots with some attributes we associate with autonomy. The term "autonomy" has so many societal connotations that it interferes with an informed understanding of artificial intelligence. The media often use autonomy to connote a robot with human-equivalent intelligence and sentience that can also act with unbounded initiative. The public perception of autonomous robots and how they should be limited is also influenced by Asimov's Three Laws of Robotics (though, as hardcore science fiction aficionados know, Asimov eventually posed four laws). Hopefully at this point the reader has discarded any notion that the Three Laws are more than a very clever plot device, along with the notion that neural networks are the only mechanism involved in achieving peer-level intelligence and autonomy.

As presented in the commentary for each chapter, artificial intelligence researchers and roboticists think of autonomy differently than what is presented in the media. The five "autonomy" stories and commentary support key insights described in the following that help us understand the limits of autonomous systems:

Autonomy is not the same as political autonomy, that is, being able to say "you're not the boss of me!" In robotics, autonomy refers to control laws that allow a robot to perform one or more capabilities automatically without a human in the loop. Robots do not set their own goals, though they may have the latitude to generate and select unexpected methods to accomplish a goal, as exemplified by Ilse in "Long Shot." In "Second Variety," the robots are doing their best to meet a poorly conceptualized goal given to them by humans. In "Runaround," the robot is trapped in a nonproductive behavior loop because of conflicting goals given by humans. Even if robots had political autonomy, if they could pick and choose what they did, Herb Simon's studies of bounded rationality in humans indicate that agents who are docile and work with others are more successful.[3] The implication of Simon's work is that really intelligent autonomous robots would not take over the world by force but instead would maximize their political autonomy and survival by working well with others. Ian M. Banks's Culture universe is an example of this in science fiction where humans and machine intelligences work together in mutual respect.

Autonomy is not the same as initiative. As described in the discussion of "Catch That Rabbit," autonomous robots may have differing levels of initiative. David in "Supertoys Last All Summer Long" has a low degree of initiative with enough process and systems-state autonomy to function but without the highest level of initiative that would enable him to relax the constraints that place him in his heartbreaking situation. Ilse in "Long Shot" exhibits the second-highest level of initiative in intentional autonomy but presumably does not have the highest level of constraint initiative needed to relax the constraint of meeting the hidden goal of restarting the human race.

More confusing, *autonomy is not the same as intelligence*. In the near future, a telepresence robot, even though it is "slaved" to an operator, may be more intelligent than an unmanned aerial vehicle (UAV) surveying a pipeline out of range of human communication and supervision. Ilse in "Long Shot" is clearly autonomous as she plans her flight path and data collection schedule, adapts the plans to environmental conditions, monitors progress, and handles any faults or failures that occur. She uses a suite of reactive and deliberative capabilities to produce her overall intelligence. Being an effective assistant and team member (which was not explored in "Stranger in Paradise" even though telepresence is a human-robot partnership), however, requires reactive and deliberative intelligence to do routine work as well as interactive intelligence, such as understanding someone else's goals, inferring their preferences and where they are likely to need assistance, adapting displays to their needs, and then generating, monitoring, selecting, and executing the best actions to support the goals. It seems unproductive to argue which system is more intelligent; instead, the metric is whether a system meets its goal.

Machine learning is not the silver bullet that will produce constraint autonomy. "Second Variety" typifies stories in which machine learning is the technobabble explanation for how robots suddenly become intelligent. "Long Shot" is more realistic. Ilse learns through experimentation how to control her complex operations, but it should be noted that she learns *how* to do things, not *what* to do. Also, what most people think of as learning is not what researchers define as machine learning. Neural networks are just one category of algorithm. For example, Ilse "learns" by using case-based reasoning, where she adapts algorithms and solutions from her libraries, rather than "learning" in the formal AI sense of using experience to improve performance. Regardless of definitions, learning is important and pervasive in autonomous robots. Whereas the primitives of SENSE, PLAN, and ACT have clear relationships in reactive and deliberative intelligence, LEARN in different forms permeates reaction, deliberation, and interaction, as well as enabling improved performance, but it doesn't alter the performance goal itself. Machine learning will just make whatever autonomous capabilities the robot has better.

Autonomy is not the same as sentience. Sentience is the ability to be self-aware, to think subjectively with a meta-awareness. It generally connotes more than meta-cognition, or forming abstractions, and incorporates expressing those subjective preferences as feelings. We have historically defined sentience as what sets us apart from, and above, animals. We can think and we care about our *ideas*, not just about our families and our immediate survival. Of course, we are constantly revising our view of humans as the only sentient species, with primates and dolphins being reconsidered for membership in this elite club; David Brin's Hugo and Nebula Award–winning Uplift universe of novels explore how the line is drawn on Earth and how it might be drawn by other galactic species. Returning to the stories in this book, the tragedy in "Supertoys Last All Summer Long" is that David has unrecognized sentience. In "Long Shot," Ilse reasons but she does so without the feelings and degree of subjectivity that color our expectations of sentience. Speedy in "Runaround" and DV-5 in "Catch That Rabbit" have much less reasoning capability than Ilse and no sentience. The varieties of robots in "Second Variety" may be so disturbing in part because they can mimic sentience without actually having it, challenging our definition of it.

Lack of trust of autonomy is believed to be primarily due to the association of autonomy with initiative combined with inappropriate testing and evaluation methods. "Trust" is a term used by the public to express fears of or discomfort with autonomous robots, most often to ask, how can we trust them? One way to gain trust in a system is to understand what it is doing and what the bounds of its operations are. As already discussed, autonomy is not the same as initiative; robots will not exceed the parameters they have been programmed with. Yet even if a user (or regulatory agency) understands that a robot will not exceed its level of initiative, they generally want proof that the robot will work as intended. The general assumption is that testing will be deterministic, that is, precisely repeatable, but deterministic methods do not work well for robots in the open world. If given the task to drive to a location, not hit anything or anyone, and follow traffic laws, a robot will execute those tasks within its programmed, finite suite of responses. The exact moment it will swerve to avoid a pedestrian or the direction it will swerve in may not be known because it will depend on the exact situation of where the robot is, where the person is, how fast they both are moving, and so on. Nondeterminism is common in computers, where computers may crash but do not create new tasks for themselves. Any "new" tasks are due to bugs or background programs that have been enabled implicitly or through an infection. Nondeterminism simply requires a different style of testing and evaluation than that for mechanical hardware. Technologists who try to use "autonomy" to excuse programming errors and sloppy testing and evaluation (e.g., "the UAV flew off on its own—it's autonomous, I had no control") are legally liable for poor engineering and may damage the market for robots by further confusing the public about autonomy.

Although existing robots are often autonomous and rarely given more than intentional initiative, this begs the question of whether there will be a singularity that endows robots with human-equivalent sentience and unbounded initiative. Certainly computing power continues to increase and on a hardware level it is beginning to approximate the theoretical information capacity of the human brain. Unfortunately, software progress lags hardware, so even though computers might have computing capacity similar to that of a brain, the software to use that capacity is further behind in development. Perhaps our current understanding of programming and emergent systems is so primitive that there will one day be a Copernican moment when the principles of bounded rationality and artificial intelligence will be upended, just like the idea that the sun revolved around the earth. But given that robots such as vacuum cleaners, autonomous cars, and unmanned aerial vehicles are becoming commercially available, we must concede that the current understanding of artificial intelligence does produce results. The success to date suggests artificial intelligence principles are on the right track, much like how Newtonian physics was correct but over time became a subset of the larger, more comprehensive field of quantum physics.

What Topics Are Missing in This Book?

These stories were chosen to facilitate discussion of core ideas of how intelligent robots are, or are not, programmed. They illustrate software architectural principles of programming intelligence and introduce aspects of a few of the major disciplines within robotics, in particular, telesystems, situated agency, behavioral robotics, potential field methodologies, deliberation, natural language understanding, theory of mind, human-robot interaction, and machine learning. They also bring up the less technical but more publicly prevalent topics of the Turing test, the uncanny valley, and ethical issues posed by robotics. On the other hand, six stories—or seven or ten or even twenty—cannot capture all of the important topics in and aspects of artificial intelligence for robotics, and thus this book serves only as a starting point.

A major topic in robotics that the stories neglect is *how do robots perceive the world?* Perception is perhaps the most diverse topic in robots as SENSE is one of the four fundamental building blocks of a robot. Robots perceive using one or more sensors paired with sensing algorithms. The algorithms are as important as the sensors, which could not function without them, just as our eyes would be useless without the brain to process light signals. Keep in mind, the processing required to see and recognize objects, simultaneously locate where we are in the world, and create internal three-dimensional maps of where we have been and what we are seeing takes up over 50 percent of the cortex of our brain.[4] It is no wonder that there is an entire field of inquiry devoted to the robot equivalent of simultaneous localization and mapping (SLAM). One problem with pundits arguing that Asimov's Three Laws of Robotics should be applied to existing robots is that the level of perception required to recognize people and situations is extremely difficult to achieve.

Robot perception is often organized around the modality or equivalent sense in humans. Each sense often has an entire discipline of researchers devoted to just that topic and roboticists draw from their findings. The largest such discipline is computer vision; it deals with perception from any sensor that produces images, including those produced at frequencies we can't see. Another large field of research is in tactile and haptic sensing, paralleling human touch. Regardless of sense, the way it is integrated into an architecture depends on the tier or layer in the hybrid deliberative/reactive organization. Sensing for behaviors relies on *direct perception* whereas sensing for deliberative or interactive capabilities relies on *object and scene recognition*. Another computing aspect of sensing is *sensor fusion*. The different tiers can use the same sensor simultaneously, processing the sensor output separately by each function, but the processes can share or fuse their perceptions as well.

Another topic we have not discussed is *how do teams of multiple robots work, especially swarms?* The study of multiple robots is often called multirobots, multiagents, or multirobot coordination. A group of robots are generally considered homogeneous, that is, identical, or heterogeneous, meaning different. "Catch That Rabbit" describes a small team that operates under the centralized control of DV-5. The robots seem to be identical except that one is given the extra compute power and authority to be the leader; they are physically homogeneous but mentally heterogeneous, so the team would be considered heterogeneous. A special class of homogeneous robots is swarms, which are large numbers of homogeneous robots with, in AI implementations, minimal reactive intelligence. Swarms work together mindlessly, emulating insects in order to accomplish tasks that can't be accomplished individually.[5] "Second Variety" is about heterogeneous robots who each operate independently, even to kill another of their own kind.

A subtle aspect of robotics that is not covered in this book is *what do robots look like?* The stories we've discusses are remarkably unimaginative in terms of robot appearance. All of the stories are about ground robots with the exception of "Long Shot," in which the robot is a spaceship. In practice, robots are not confined to the ground. Unmanned aerial vehicles and unmanned marine vehicles are becoming commonplace and offer exciting new applications of robotics for search and rescue, environmental protection, and critical infrastructure inspection.

The ground robots in the stories also look either humanoid or nearly so, but commercial robots are rarely human-like. Conceptualizing robots as humanoid may produce designs that are less effective. Ground robots already take many forms, be it miniature tanks, cars, snakes, cockroaches, dogs, and even caterpillars because in engineering, form follows function. Robots typically do things that humans cannot, so there is no incentive to design them by default to look like humans. And as per the discussion of the uncanny valley in "Supertoys Last All Summer Long," if a humanoid robot is used, the designer has to work very hard to avoid making it creepy or distasteful.

Further Reading

Hopefully these six stories and commentary inspire a desire to read and learn more.

For a more detailed exposition of autonomy and artificial intelligence that is still targeted for a general audience, the 2012 Defense Science Board Task Force report on the role of autonomy for the Department of Defense is probably the best resource.[6] The study provides a concise statement on how unmanned systems (robots) have been used by the US Department of Defense and the technological advances required to use them more effectively and ethically in the future. It specifically describes the state of the practice in perceptual processing, planning, learning, human-robot interaction, natural language understanding, and multiagent coordination, as well as technical gaps. As full disclosure, I cochaired the study so I am biased, but I believe that it captures the present and future of AI and robotics better than any other broad government study I have participated in over the past twenty years. The 2012 study is not the same as the 2015 study, which I also served on. The 2015 study is much broader in scope, and therefore I recommend the shorter 2012 study for an assessment of the state of artificial intelligence for robots.

For exploring specific algorithms for artificial intelligence for robotics, I recommend two books. One is my *Introduction to AI Robotics* for a survey of classes of algorithms (*Robotics Through Science Fiction* was conceived of in part as a companion to the second edition of this book), and the other is the *Springer Handbook of Robotics* for details on specific algorithms.[7] The *Handbook* is formidable in size and cost, but it has won multiple awards for the clarity of its writing. It should be understandable to someone with a background in engineering or computer science and gives pointers to the hardcore scientific papers in each area. When perusing the *Handbook*, keep in mind that "robotics" does not necessarily mean "artificial intelligence for robots"—for example, many factory robot arms are mechanical automata and famously mindless. Therefore there are many chapters that may not be of interest. The *Springer Handbook of Robotics* is best treated as an encyclopedia rather than something to be read beginning to end. Also note that introductory textbooks on artificial intelligence may be too general to be of much benefit, as "artificial intelligence" does not necessarily mean "artificial intelligence for robots."

And finally, as a professor, I cannot resist recommending taking classes in artificial intelligence and robotics. There are many classes at universities and from different web delivery systems, including mine. Do keep in mind that artificial intelligence for robotics is like business and economics: there are competing schools of thought. Just as there are schools of business and economics represented by Harvard, the University of Chicago, the Wharton School, Stanford, and so on, there are flavors of artificial intelligence for robotics and thus there exists a rich, and sometimes contentious, diversity of approaches and classes. One class may not be enough and a particular school of thought may not resonate with all learning styles or academic backgrounds. Don't let a "meh" experience with one class dissuade life-long learning about a core technology for the future.

Notes

Introduction

1. David Kortenkamp, R. Peter Bonasso, and Robin Murphy, eds., *Artificial Intelligence and Mobile Robots: Case Studies of Successful Systems* (Menlo Park, CA: AAAI Press, 1998).

2. Michael Crichton, *Prey* (New York: HarperCollins, 2002).

3. Horace Miner, "Body Ritual Among the Nacirema," Reprint Series in Social Sciences (New York: Irvington Publishers, 1956 and 1993); and Carol Mason, Martin Harry Greenberg, and Patricia Warrick, eds., *Anthropology Through Science Fiction* (New York: St. Martin's Press, 1974).

4. Robert A. Heinlein, *The Green Hills of Earth* (Riverdale, NY: Baen Books, 1951).

5. Karel Čapek, *R.U.R.*, trans. David Wyllie (Rockville, MD: Wildside Press, 2010).

Chapter 1

1. Isaac Asimov, "Stranger in Paradise," in *The Complete Robot* (New York: Doubleday, 1982).

2. Thomas Sheridan, *Telerobotics, Automation, and Human Supervisory Control* (Cambridge, MA: MIT Press, 1992).

3. Robin R. Murphy, *Disaster Robotics* (Cambridge, MA: MIT Press, 2014).

4. Robin R. Murphy, *Introduction to AI Robotics*, 2nd ed. (Cambridge, MA: MIT Press, 2018).

5. Sheridan, *Telerobotics, Automation, and Human Supervisory Control.*

6. C. Wampler, "Teleoperators, Supervisory Control," in *Concise International Encyclopedia of Robotics: Applications and Automation*, ed. Richard C. Dorf and Shimon Y. Nof (New York: Wiley-Interscience, 1990), 997.

7. Murphy, *Introduction to AI Robotics.*

8. N. J. Nilsson, *Shakey the Robot* (Menlo Park, CA: Sri International, 1984).

Chapter 2

1. Isaac Asimov, "Runaround," in *The Complete Robot* (New York: Doubleday, 1982).

2. Robin R. Murphy, *Introduction to AI Robotics*, 2nd ed. (Cambridge, MA: MIT Press, 2018); and Ronald C. Arkin, *Behavior-based Robotics* (Cambridge, MA: MIT Press, 1998).

3. N. J. Nilsson, *Shakey the Robot* (Menlo Park, CA: Sri International, 1984).

4. Ronald C. Arkin, *Behavior-Based Robotics* (Cambridge, MA: MIT Press, 1998).

5. J. J. Gibson, *The Ecological Approach to Visual Perception* (Boston: Houghton Mifflin, 1979).

6. Michael A. Arbib and Jim-Shih Liaw, "Sensorimotor Transformations in the Worlds of Frogs and Robots," *Artificial Intelligence* 72, no. 1–2 (1995): 53–79.

Chapter 3

1. Vernor Vinge, "Long Shot," first published in *Analog Science Fiction/Science Fact*, ed. Ben Bova (New York: Conde Nast, 1972).

2. Hans Moravec, *Mind Children: The Future of Robot and Human Intelligence* (1988; repr., Cambridge, MA: Harvard University Press, 1990).

3. Vernor Vinge, *The Collected Stories of Vernor Vinge* (New York: Tor Books, 2001; repr. New York: Orb Books, 2002).

4. Brian C. Williams, Michel D. Ingham, Seung Chung, Paul Elliott, Michael Hofbaur, and Gregory T. Sullivan, "Model-Based Programming of Fault-Aware Systems," *AI Magazine* 24, no. 4 (2003): 61–75.

5. Robin R. Murphy, *Introduction to AI Robotics*, 2nd ed. (Cambridge, MA: MIT Press, 2018).

6. Theodore Sturgeon, "The Man Who Lost the Sea," in *The Man Who Lost the Sea: Volume X; The Complete Stories of Theodore Sturgeon* (Berkeley, CA: North Atlantic Books, 2005).

7. A. Newell, J. C. Shaw, and H. A. Simon, "Report on a General Problem-Solving Program," *Information Processing: Proceedings of the International Conference on Information Processing, Unesco, Paris, 15–20 June 1959* (Paris: Unesco, 1960), 256–264.

8. Williams et al., "Model-Based Programming."

9. J. S. Albus, "RCS: A Reference Model Architecture for Intelligent Control," *Computer* 25, no. 5 (1992): 56–59.

Chapter 4

1. Isaac Asimov, "Catch That Rabbit," in *The Complete Robot* (New York: Doubleday, 1982).

2. Robin R. Murphy, *Introduction to AI Robotics*, 2nd ed. (Cambridge, MA: MIT Press, 2018).

3. Tucker Balch and Lynne E. Parker, eds., *Robot Teams: From Diversity to Polymorphism* (Natick, MA: A K Peters, 2002).

4. Defense Science Board, *The Role of Autonomy in DoD Systems*, DTIC ADA566864 (Washington, DC: Department of Defense, 2012).

5. Randall Davis and Douglas Lenat, *Knowledge-Based Systems in Artificial Intelligence* (New York: McGraw-Hill, 1981).

6. Alan Colman and Jun Han, "Roles, Players and Adaptable Organizations," *Applied Ontology* 2, no. 2 (April 2007): 105–126.

7. Bruno Siciliano and Oussama Khatib, eds., *Springer Handbook of Robotics*, 2nd ed. (Berlin Heidelberg: Springer-Verlag, 2016).

Chapter 5

1. Brian Aldiss, "Supertoys Last All Summer Long," *Harper's Bazaar*, December 1969.

2. Robin R. Murphy, *Introduction to AI Robotics*, 2nd ed. (Cambridge, MA: MIT Press, 2018).

3. Stuart Russell and Peter Norvig, *Artificial Intelligence: A Modern Approach*, 3rd ed. (Harlow, Essex, UK: Pearson, 2009).

4. Masahiro Mori, "The Uncanny Valley: The Original Essay by Masahiro Mori," trans. Karl F. MacDorman and Norri Kageki, June 12, 2012, https://spectrum.ieee.org/automaton/robotics/humanoids/the-uncanny-valley.

5. Carolyn Gregoire, "What This Robotic Baby Can Teach Us about How Infants Communicate," Huffington Post, September 25, 2015, last updated September 28, 2015, http://www.huffingtonpost.com/entry/robotic-baby-infants-moms_56041996e4b0fde8b0d18334.

6. Michael Georgeff, Barney Pell, Martha Pollack, Milind Tambe, and Michael Wooldridge, "The Belief-Desire-Intention Model of Agency," in *International Workshop on Agent Theories, Architectures, and Languages* (London: Springer-Verlag, 1998), 1–10.

7. Clifford Nass, "Ecce Homo: Why It's Great to Be Labeled a 'Person'" (PowerPoint presentation, DARPA/NSF Workshop on Human-Robot Interaction, San Luis Obispo, CA, September 29–30, 2001), www.erogersphd.com/EMorePages/HRI/HRI-ARCHIVE/cnass.ppt.

Chapter 6

1. Philip K. Dick, "Second Variety," first published in *Space Science Fiction* 1, no. 6 (May 1953); now available at http://www.gutenberg.org/files/32032/32032-h/32032-h.htm.

2. "UN Meeting Targets 'Killer Robots,'" UN News, May 14, 2014, https://news.un.org/en/story/2014/05/468302-un-meeting-targets-killer-robots#.VjdQw-doxlq.

3. Ronald Arkin, *Governing Lethal Behavior in Autonomous Robots* (Boca Raton, FL: CRC Press, 2009).

4. Karel Čapek, *R.U.R.*, trans. David Wyllie (Rockville, MD: Wildside Press, 2010).

5. See https://en.wikipedia.org/wiki/With_Folded_Hands.

6. See https://en.wikipedia.org/wiki/Self-replicating_machines_in_fiction.

7. Robin R. Murphy, *Introduction to AI Robotics*, 2nd ed. (Cambridge, MA: MIT Press, 2018).

8. W. Wallach and Colin Allen, *Moral Machines: Teaching Robots Right from Wrong* (Oxford: Oxford University Press, 2010).

9. Marge Piercy, *He, She and It* (New York: Fawcett, 1993).

10. See https://en.wikipedia.org/wiki/Samsung_SGR-A1.

11. Erik Sofge, "Top 5 Bomb-Packing, Gun-Toting War Bots the U.S. Doesn't Have," Popular Mechanics, September 30, 2009, http://www.popularmechanics.com/military/a5439/4249209/.

12. Robin R. Murphy and David D. Woods, "Beyond Asimov: The Three Laws of Responsible Robotics," *IEEE Intelligent Systems* 24, no. 4 (2009): 14–20.

13. Herbert A. Simon, *The Sciences of the Artificial*, 3rd ed. (Cambridge, MA: The MIT Press, 1996).

14. Stuart Russell and Peter Norvig, *Artificial Intelligence: A Modern Approach*, 3rd ed. (Upper Saddle River, NJ: Pearson, 2009).

Chapter 7

1. N. J. Nilsson, *Shakey the Robot* (Menlo Park, CA: Sri International, 1984).

2. Ulric Neisser, *Cognition and Reality—Principles and Implications of Cognitive Psychology* (New York: W. H. Freeman and Company, 1976).

3. Herbert A. Simon, *The Sciences of the Artificial*, 3rd ed. (Cambridge, MA: The MIT Press, 1996).

4. Susan Hagen, "The Mind's Eye," *Rochester Review* 74, no. 4 (2012): 32–37.

5. Tucker Balch and Lynne E. Parker, eds., *Robot Teams: From Diversity to Polymorphism* (Natick, MA: A K Peters, 2002).

6. Defense Science Board, *The Role of Autonomy in DoD Systems*, DTIC ADA566864 (Washington, DC: Department of Defense, 2012).

7. Robin R. Murphy, *Introduction to AI Robotics*, 2nd ed. (Cambridge, MA: MIT Press, 2018); and Bruno Siciliano and Oussama Khatib, eds., *Springer Handbook of Robotics*, 2nd ed. (Berlin Heidelberg: Springer-Verlag, 2016).

Index